Flutter Design Patterns and Best Practices

Build scalable, maintainable, and production-ready apps using effective architectural principles

Daria Orlova

Esra Kadah

Jaime Blasco

Flutter Design Patterns and Best Practices

Copyright © 2024 Packt Publishing

Group Product Manager: Rohit Rajkumar

Publishing Product Manager: Kaustubh Manglurkar

Book Project Manager: Sonam Pandey

Senior Editor: David Sugarman

Technical Editor: K Bimala Singha

Copy Editor: Safis Editing

Proofreader: David Sugarman

Indexer: Pratik Shirodkar

Production Designer: Vijay Kamble

DevRel Marketing Coordinators: Anamika Singh and Nivedita Pandey

Publication date: September 2024

Production reference: 2290126

Published by Packt Publishing Ltd.

Grosvenor House

11 St Paul's Square

Birmingham

B3 1RB, UK

ISBN 978-1-80107-264-9

www.packtpub.com

To the memory of my grandmothers, Frosya and Ira, who were my guiding stars during their time and even more so after. To my husband, Alex, for his infinite support in everything I do. To my little brother, Matvey, who inspires me to be my best. And to my parents, Sergey and Tatyana, for raising me to be the person I am today.

– Daria Orlova

To my beloved parents, Emine and Erdinc, who have been the foundation of my journey, always striving to provide the best for me and grounding me with their dedicated support.

To my dear sisters, Ceren and Neslihan, and my brother, Enes, whose encouragement and companionship have guided me through good times and bad. Your patience, countless hours of listening, and unwavering support have profoundly influenced every decision I have made.

To my amazing professors, Gurkan Ozturk and Emre Cimen from Eskisehir Technical University, who introduced me to coding and consistently supported and motivated me. Your guidance made sure I never felt alone as I started on this path, planting the seeds for my future success.

– Esra Kadah

Dedicated to the memory of my father, whose love was my constant source of strength. Your belief in me knew no bounds, and I am eternally grateful for your support.

– Jaime Blasco

Contributors

About the authors

Daria Orlova is a mobile app specialist, who started with native Android in 2015 and discovered Flutter in 2019. She is the co-founder of **BBapps**, creating *"apps that care for you and the planet."* Previously, at Chili Labs, the top mobile agency in the Baltics, together with the team, she delivered 50+ projects, winning awards including RedDot and W3. Daria is an active community member – a **Google Developer Expert** (**GDE**) in Flutter and Dart, Flutterista, and WTM ambassador, mentor, technical writer, and public speaker who is passionate about quality tech education. She holds a B.Sc. in computer science.

I want to thank my friends, Anna Leushchenko, Cagatay Ulusoy, Dmitry Zhifarsky, and Tair Rzayev, whose expertise has greatly helped to make this book better. Thank you to Packt and tech reviewer Ahmed Fathy for making this book happen, and to Chili Labs, for providing the experience that was put into the foundation of this book.

Esra Kadah is a senior app developer specializing in Flutter, with a passion for integrating programming, UI/UX design, and psychology. She enjoys building and contributing to thriving communities. She is a co-organizer of Flutteristas, Flutter Berlin, and CMX Istanbul, and serves as a Women Techmakers Ambassador. She has delivered over 60 talks, organized more than 150 events, and hosted 40+ streams, collaborating with Google Developer Groups, Google Developer Student Clubs, the Flutter community, Flutteristas, and Women Techmakers.

Jaime Blasco is a seasoned Flutter developer with a deep-rooted passion for crafting exceptional mobile experiences. As a **Google Developer Expert** (**GDE**) in Flutter, he possesses an in-depth understanding of the framework and its capabilities. His active involvement in the Flutter community, notably as a co-leader of the Spanish Flutter community, underscores his commitment to knowledge sharing and collaboration. He currently develops the Hypervolt app, a Flutter-based solution that seamlessly connects users to EV chargers. Jaime ensures smooth user interactions and efficient hardware communication, delivering a reliable charging experience.

About the reviewer

Ahmed Fathy is a skilled senior software engineer from Alexandria, Egypt. He holds a B.Sc. in computer and systems engineering from Al-Azhar University. Specializing in Flutter development, he has extensive experience as an instructor with private training companies and his own online courses and in projects held by Egyptian ministries of communication.

Ahmed has developed numerous mobile applications. He is a community leader at GDG Alexandria and a previous co-founder of Google DSC Al-Azhar. Ahmed is passionate about civil work, community building, and sharing knowledge.

Table of Contents

Part 1: Building Delightful User Interfaces

1

2

Part 2: Connecting UI with Business Logic

3

4

5

Creating Consistent Navigation   151

Part 3: Exploring Practical Design Patterns and Architecture Best Practices

6

The Responsible Repository Pattern   175

7

Implementing the Inversion of Control Principle   201

8

Ensuring Scalability and Maintainability with Layered Architecture 215

9

Mastering Concurrent Programming in Dart   233

10

A Bridge to the Native Side of Development   259

Part 4: Ensuring App Quality and Stability

11

Unit Tests, Widget Tests, and Mocking Dependencies   293

Preface

Messaging friends, booking airplane tickets, ordering a grocery delivery, checking bank accounts, buying metro tickets... this is just a short list of tasks we accomplish today with the help of mobile and web applications. Apps are omnipresent, and someone must develop them. If you are holding this book, there is a high chance that you are one of these developers.

In recent years, Flutter has become a stable and widely used framework for building apps. And not just mobile apps, as it also supports building for web, desktop, and beyond. However, to thrive in the modern world, apps need to be more than just functional – they must be beautiful, fast, and reliable. These qualities are achieved through the approaches used to build the apps. A scalable, flexible, maintainable, and testable architecture is essential to help businesses stand out and to provide users with the high-quality experience they expect. This is where design patterns and best practices come into play.

Design patterns are proven blueprints for solutions to common problems that arise in software design. They provide standard terminology and are specific to particular scenarios, making the development process more efficient and reliable. For instance, patterns such as Singleton, Observer, and Factory Method offer templates for solving issues related to object creation, communication between objects, and more.

Best practices, on the other hand, are guidelines or methodologies that have been shown through experience and research to lead to optimal results. These practices include coding standards, architectural principles, and development processes that ensure high-quality software. They help to maintain code readability, performance, security, and scalability.

This book dives into the details of Flutter's inner workings, teaches various design patterns to build Flutter apps, and explores the best practices for developing robust applications. Understanding these fundamentals is crucial for making informed decisions about which practices and guidelines to follow and which to adapt or skip.

This knowledge is built on the experience of developing over 50 apps of various scales in industry-leading mobile development agencies and companies. However, it's important to remember that there is always room for individual opinions and adjustments.

It is a great time to be a Flutter developer, and this book will help you become one who is highly skilled and competitive.

Who this book is for

Mobile developers of any level can gain practical insights into how to craft Flutter apps according to best practices. You are especially likely to benefit from this book if you belong to one of the following groups:

- **Flutter developers**: If you have already built some projects with Flutter and want to enhance your skills to build scalable, maintainable, and stable applications that follow the best practices, this book will show you how.

- **Mobile developers from other tech stacks**: If you have already built mobile apps in other frameworks, such as React Native or Xamarin, or for native platforms, and want to transition to Flutter, this book will teach you how to apply your existing knowledge to Flutter.

- **Aspiring Flutter developers**: If you have not yet built apps in any tech stacks but have some programming experience in other stacks, this book can be used to navigate the Flutter framework alongside more beginner-friendly resources.

What this book covers

Chapter 1, *Best Practices for Building UIs with Flutter*, discusses the difference between imperative and declarative approaches to UI building and why modern frameworks prefer the latter. We will explore the details of the Flutter widget tree system, and practical advice on how to build performant interfaces.

Chapter 2, *Responsive UI for All Devices*, provides an overview of the Flutter layout algorithm, dives into the best practices and available options for building responsive interfaces, and covers accessibility best practices.

Chapter 3, *Vanilla State Management*, opens the topic of state management in Flutter. It provides the definition of state and its different types. In this chapter, we start building the Candy Store app, which we will continue building throughout the book, and learn how to implement state management patterns in the vanilla Flutter way. You will also see an overview of the `InheritedWidget` class details, and practical tips on working with `BuildContext` in Flutter.

Chapter 4, *State Management Patterns and Their Implementations*, continues the topic of state management, introducing popular industry patterns such as MVVM and MVI, the rationale behind using them, and their implementation in Flutter with and without third-party libraries.

Chapter 5, *Creating Consistent Navigation*, provides an overview of navigation patterns in Flutter, going into details on how to implement imperative style navigation and declarative style navigation. We will see some examples of building complex navigation scenarios and when to choose which approach.

Chapter 6, *The Responsible Repository Pattern*, introduces the Repository pattern and its benefits for scalable app architecture. The chapter goes deep into implementation details and explores practices for building flexible data sources.

Chapter 7, Implementing the Inversion of Control Principle, explores various approaches to implementing the Inversion of Control principle, via practices such as dependency injection and the Service Locator pattern, and demonstrates their practical application with the help of different libraries.

Chapter 8, Ensuring Scalability and Maintainability with Layered Architecture, provides an overview of how to structure the code that we have built up to this point according to layered architecture principles. The chapter also highlights how we have been following the SOLID and other best software design principles all along.

Chapter 9, Mastering Concurrent Programming in Dart, introduces concepts related to concurrent programming in general and provides an overview of asynchronous APIs in Dart. The chapter goes into details of how to work efficiently with the Future APIs, as well as how to handle parallel operations with the Isolates API.

Chapter 10, A Bridge to the Native Side of Development, provides an overview of the Flutter app architecture from the perspective of the SDK and hosting platforms. The chapter goes into details of working with platform channels, a mechanism used to communicate with the host platform, as well as demonstrating the shortcomings of this API. We then explore a type-safe way to implement that communication via the `pigeon` code generation library.

Chapter 11, Unit Tests, Widget Tests, and Mocking Dependencies, provides an overview of the automated testing approaches in Flutter. You will learn how to write unit tests and go into the details of widget testing. The chapter showcases the mocking dependencies technique and how to implement it with the Mockito library.

Chapter 12, Static Code Analysis and Debugging Tools, discusses the topic of static analysis in Flutter and why it's important to establish coding conventions and follow them consistently. We then see how this can be automated by setting up a robust static analysis system. The chapter also explores debugging practices and their applications, such as logging, assertions, breakpoints, and Flutter DevTools.

To get the most out of this book

You will need to download an IDE that supports development with Flutter and Dart, and the Flutter SDK itself.

Software covered in the book	Operating system requirements
Flutter SDK 3.22.0+	Windows, macOS, Linux, or ChromeOS
Dart 3.4.0+	Windows, macOS, Linux, or ChromeOS

You may use any IDE of your choice, but some popular ones that support Flutter are Android Studio, VS Code, and IntelliJ IDEA. Up-to-date details for installation can always be viewed at the official website: `https://docs.flutter.dev/get-started/install`.

If you are using the digital version of this book, we advise you to type the code yourself or access the code from the book's GitHub repository (a link is available in the next section). Doing so will help you avoid any potential errors related to the copying and pasting of code.

The best way to read this book is consecutively, chapter by chapter, as we start with the basics and build on top of the previous chapter with every step. That said, you may still find individual chapters useful if you're searching for specific topics – just remember they are part of the bigger project.

Download the example code files

You can download the example code files for this book from GitHub at `https://github.com/PacktPublishing/Flutter-Design-Patterns-and-Best-Practices`. If there's an update to the code, it will be updated in the GitHub repository.

We also have other code bundles from our rich catalog of books and videos available at `https://github.com/PacktPublishing/`. Check them out!

Conventions used

There are a number of text conventions used throughout this book.

`Code in text`: Indicates code words in text, database table names, folder names, filenames, file extensions, pathnames, dummy URLs, user input, and Twitter handles. Here is an example: "Here's how we could use the `Align` widget to achieve the same effect."

A block of code is set as follows:

```
Container(
  constraints: BoxConstraints.tight(
    const Size(200, 100),
  ),
  color: Colors.red,
  child: const Text('Hello World'),
);
```

When we wish to draw your attention to a particular part of a code block, the relevant lines or items are set in bold:

```
Container(
  alignment: Alignment.center,
  constraints: BoxConstraints.tight(
    const Size(200, 100),
  ),
  color: Colors.red,
  child: const Text('Hello World'),
);
```

Bold: Indicates a new term, an important word, or words that you see onscreen. For instance, words in menus or dialog boxes appear in **bold**. Here is an example: "Select **System info** from the **Administration** panel."

> **Tips or important notes**
> Appear like this.

Get in touch

Feedback from our readers is always welcome.

General feedback: If you have questions about any aspect of this book, email us at customercare@packtpub.com and mention the book title in the subject of your message.

Errata: Although we have taken every care to ensure the accuracy of our content, mistakes do happen. If you have found a mistake in this book, we would be grateful if you would report this to us. Please visit www.packtpub.com/support/errata and fill in the form.

Piracy: If you come across any illegal copies of our works in any form on the internet, we would be grateful if you would provide us with the location address or website name. Please contact us at copyright@packt.com with a link to the material.

If you are interested in becoming an author: If there is a topic that you have expertise in and you are interested in either writing or contributing to a book, please visit authors.packtpub.com.

Free benefits with your book

This book comes with free benefits to support your learning. Activate them now for instant access (see the "*How to Unlock*" section for instructions).

Here's a quick overview of what you can instantly unlock with your purchase:

DRM-Free PDF Version

Download DRM-free PDF and ePub copies of this book.

7-Day Packt Library Access

Get 7-day unlimited access to 8,000+ books and videos. No credit card required.

Available for first-time Packt+ trial users only.

Next-Gen Reader Access

Read this book on Packt Reader with progress sync, dark mode and note-taking.

How to Unlock

Scan the QR code (or go to `packtpub.com/unlock`). Search for this book by name, confirm the edition, and then follow the steps on the page.

Note: Keep your invoice handy. Purchases made directly from Packt don't require one

Share Your Thoughts

Once you've read *Flutter Design Patterns and Best Practices*, we'd love to hear your thoughts! Scan the QR code below to go straight to the Amazon review page for this book and share your feedback.

`https://packt.link/r/1-801-07264-7`

Your review is important to us and the tech community and will help us make sure we're delivering excellent quality content.

Part 1:
Building Delightful
User Interfaces

In this part, you will learn how to build beautiful **user interfaces** (**UIs**) with Flutter and how to do that in a productive and efficient manner. The topics that we will cover include the difference between imperative and declarative UI paradigms, the details of the inner workings of the most important concept in Flutter – widgets, best practices for working with widgets and their lifecycle, the Flutter layout algorithm, and various techniques for building responsive and accessible UIs.

This part includes the following chapters:

- *Chapter 1, Best Practices for Building UIs with Flutter*
- *Chapter 2, Responsive UIs for All Devices*

1

Best Practices for Building UIs with Flutter

Flutter is rapidly becoming a go-to framework for creating applications of various scales. Google Trends and Stack Overflow confirm that Flutter has become a more popular search term than React Native for the last several years (see `https://trends.google.com/trends/explore?q=%2Fg%2F11f03_rzbg,%2Fg%2F11h03gfxy9&hl=en` and `https://insights.stackoverflow.com/trends?tags=flutter%2Creact-native`). Flutter consistently appears in various development ratings: the top 3 GitHub repositories by number of contributors (`https://octoverse.github.com/2022/state-of-open-source`), the top 3 most downloaded plugins in JetBrains Marketplace (`https://blog.jetbrains.com/platform/2024/01/jetbrains-marketplace-highlights-of-2023-major-updates-community-news/`), and second in Google Play Store ratings right after Kotlin (`https://appfigures.com/top-sdks/development/apps`).

This isn't a surprise, since Flutter offers a delightful toolkit that allows developers to build smooth and pixel-perfect UIs almost immediately after their first encounter with the framework. Flutter also does a great job of hiding away the details of the rendering process. However, because it is so easy to overlook those details, a lack of understanding of how the framework actually works can lead to performance issues.

This chapter explores the benefits of using Flutter's declarative UI-building approach, as well as how that approach affects developers. We will discuss methods for optimizing performance and avoiding interference with the framework's building and rendering processes. We will also examine how this approach works under the hood and provide best practices for creating beautiful and blazing-fast interfaces.

By the end of this chapter, you will understand the concept of the Flutter tree system and how to scope your widget tree for the best performance. This knowledge will provide the foundation necessary for learning architectural design patterns based on the framework's build system.

In this chapter, we're going to cover the following main topics:

- Understanding the difference between declarative and imperative UI design.
- Everything is a widget! Or is it?
- Reduce, reuse, recycle!

Understanding the difference between declarative and imperative UI design

The beauty of technology is that it evolves with time based on feedback about developer experience. Today, if you're in mobile development, there is a high chance that you have heard about Jetpack Compose, SwiftUI, React Native, and of course Flutter. The thing these technologies have in common is both that they're used for creating mobile applications and the fact that they do it via a declarative programming approach. You may have heard this term before, but what does it actually mean and why is it important?

To take full advantage of a framework, it's important to understand its paradigm and work with it rather than against it. Understanding the "why" behind the architectural decisions makes it much easier to understand the "how," and to apply design patterns that complement the overall system.

Native mobile platforms have a long history of development and major transitions. In 2014, Apple announced a new language, Swift, that would replace the current Objective-C. In 2017 the Android team made Kotlin the official language for Android development, which would gradually replace Java. Those introductions had a hugely positive impact on the developer experience, yet they still had to embrace the legacy of existing framework patterns and architecture. In 2019, Google announced Jetpack Compose and Apple announced SwiftUI – completely new toolkits for building UIs. Both SwiftUI and Jetpack Compose take advantage of their respective languages, Swift and Kotlin, leaving legacy approaches behind. Both toolkits also loudly boast their declarative programming paradigm. But language advantages aside, let's explore why declarative is now the industrial de facto and what is wrong with imperative.

Understanding the imperative paradigm

By definition, the imperative programming paradigm focuses on how to achieve the desired result. You describe the process step by step and have complete control of the process. For example, it could result in code such as this:

```
fun setErrorState(errorText: String) {
    val textView = findViewById<TextView>(R.id.error_text_view)
    textView.text = errorText
    textView.setTextColor(Color.RED)
    textView.visibility = View.VISIBLE
```

```
    val button = findViewById<Button>(R.id.submit_button)
    button.isEnabled = true
    val progressView = findViewById<ProgressBar>(R.id.progress_view)
    progressView.visibility = View.GONE
}
```

In the preceding snippet, we imperatively described how to update the UI in case of an error. We accessed the UI elements step by step and mutated their fields.

This is a real example of code that could've been written for a native Android application. Even though this approach may be powerful and gives the developer fine-grained control over the flow of the logic, it comes with the possibility of the following problems:

- The more elements that can change their presentation based on a state change, the more mutations you need to handle. You can easily imagine how this simple setErrorState becomes cumbersome as more fields need to be hidden or changed. The approach also assumes that there are similar methods for handling a progress and success state. Code such as this may easily become hard to manage, especially as the amount of views in your app grows and the state becomes more complex.

- Modifying the global state can produce side effects. On every such change, we mutate the same UI element and possibly call other methods that also mutate the same elements. The resulting myriad of nested conditionals can quickly lead to inconsistency and illegal states in the final view that the user sees. Such bugs tend to manifest only when certain conditions are met, which makes them even harder to reproduce and debug.

For many years, the imperative approach was the only way to go. Thankfully, native mobile frameworks have since started adopting declarative toolkits. Although these are great, developers who need to switch between paradigms inside of one project can encounter many challenges. Different tools require different skills and in order to be productive, the developer needs to be experienced with both. More attention needs to be paid to make sure that the application that is created with various approaches is consistent. While the new toolkits are in the process of wider adoption, some time and effort are required until they are able to fully implement what their predecessors already have. Thankfully, Flutter embraced declarative from the start.

Understanding the declarative paradigm

In an imperative approach, the focus is on the "how." However, in the declarative approach, the focus is on the "what." The developer describes the desired outcome, and the framework takes care of the implementation details. Since the details are abstracted by the framework, the developer has less control and has to conform to more rules. Yet the benefit of this is the elimination of the problems imposed by the imperative approach, such as excessive code and possible side effects. Let's take a look at the following example:

```
Widget build(BuildContext context) {
    final isError = false;
    final isProgress = true;
    return Column(
     children: [
       MyContentView(
         showError: isError,
       ),
       Visibility(
         visible: isProgress,
         child: Center(
           child: CircularProgressIndicator(),
         ),
       ),
     ],
    );
}
```

In the preceding code, we have built a UI as a reaction to state changes (such as the `isError` or `isProgress` fields). In the upcoming chapters, you will learn how to elegantly handle the state, but for now, you only need to understand the concept.

This approach can also be called reactive, since the widget tree updates itself as a reaction to a change of state.

Does Flutter use the declarative or imperative paradigm?

It is important to understand that Flutter is a complex framework. Conforming to just one programming paradigm wouldn't be practical, since it would make a lot of things harder (see `https://docs.flutter.dev/resources/faq#what-programming-paradigm-does-flutters-framework-use`). For example, a purely declarative approach with its natural nesting of code would, make describing a `Container` or `Chip` widget unreadable. It would also make it more complicated to manage all of their states.

Here's an excerpt from the `build` method of the `Container` describing how to build the child widget imperatively:

```
@override
Widget build(BuildContext context) {
  Widget? current = child;
  // ...
```

```
  if (color != null) {
    current = ColoredBox(color: color!, child: current);
  }
  if (margin != null) {
    current = Padding(padding: margin!, child: current);
  }
  // ...
}
```

Even though the main approach of describing the widget tree can be viewed as declarative, imperative programming can be used when it feels less awkward to do so. This is why understanding the concepts, patterns, and paradigms is crucial to creating the most efficient, maintainable, and scalable solutions.

If you are coming from an imperative background, getting used to the declarative approach of building the UI may be mind-bending at first. However, shifting your focus from "how" to "what" you're trying to build will help. Flutter can help you too, as instead of mutating each part of the UI separately, Flutter rebuilds the entire widget tree as a reaction to state changes. Yet the framework still maintains snappy performance, and developers usually don't need to think about it much.

In the next section, let's take a closer look at the abstraction to understand *how* the *what* actually works. We will explore not only how to use the widgets as a developer but also how the framework efficiently handles them under the hood. We will cover what to do and what not to do to avoid interfering with the building and rendering processes.

Everything is a widget! Or is it?

You have probably heard this phrase many times. It has become the slogan of Flutter – in Flutter, everything is a widget! But what is a widget and how true is this saying? At first glance, the answer might seem simple: a widget is a basic building block of UI, and everything you see on the screen is a widget. While this is true, these statements don't provide much insight into the internals of a widget.

The framework does a good job of abstracting those details away from the developer. However, as your app grows in size and complexity, if you don't follow best performance practices, you may start encountering issues related to frame drop. Before this can happen, let's learn about the Flutter build system and how to make the most of it.

What is a widget?

For most of our development, we will create widgets that extend `StatelessWidget` or `StatefulWidget`. The following is the code for these:

```
abstract class StatelessWidget extends Widget {...}
abstract class StatefulWidget extends Widget {...}
```

```
@immutable
abstract class Widget {...}
```

From the source code, we can see that both of these widgets are abstract classes and that they inherit from the same class: the `Widget` class.

Another important place where we see the `Widget` class is in our `build` method:

```
Widget build(BuildContext context) {...}
```

This is probably the most overridden method in a Flutter application. We override it every time we create a new widget and we know that this is the method that gets called to render the UI. But how often is this method called? First of all, it can be called whenever the UI needs an update, either by the developer, for example, via `setState`, or by the framework, for example, on an animation ticker. Ultimately, it can be called as many times as your device can render frames in a second, which is represented by the refresh rate of your device. It usually ranges from 60 Hz to 120 Hz. This means that the `build` method can be called 60-120 times per second, which gives it 16-8 ms (1,000 ms / 60 frames-per-second = 16 ms or 1,000 ms / 120 frames-per-second = 8 ms) to render the whole `build` method of your app. If you fail to do that, this will result in a frame drop, which might mean UI jank for the user. Usually, this doesn't make users happy! But all developer performance optimizations aside, surely this can't be what's happening? Redrawing the whole application widget tree on every frame would certainly impact performance. This is not what happens in reality, so let's find out how Flutter solves this problem.

When we look at the `Widget` class signature, we see that it is marked with an `@immutable` annotation. From a programming perspective, this means that all of the fields of this class have to be `final`. So after you create an instance of this class, you can't mutate any of its fields (collections are different but let's ignore this for now and return to it in *Chapter 4*). This is an interesting fact when you remember that the return type of our `build` method is `Widget` and that this method can be called up to 120 times per second. Does that mean that every time we call the `build` method, we will return a completely new tree of widgets? All million of them? Well, yes and no. Depending on how you build your widget tree and why and where it was updated, either the whole tree or only parts of it get rebuilt. But widgets are cheap to build. They barely have any logic and mostly serve as a data class for another Flutter tree that we will soon observe. Before we move on to this tree though, let's take a look at one special type of widget.

Getting to know the RenderObjectWidget and its children

We have already discussed that when dealing with widgets, we mostly extend `StatelessWidget` and `StatefulWidget`. Inside the `build` method of our widgets, we only compose them like Lego bricks using the widgets already provided by the Flutter framework, such as `Container` and `TextFormField`, or our own widgets.

Most of the time, we only use the `build` method. Less often, we may use other methods such as `didChangeDependencies` or `didUpdateWidget` from the `State` object. Sometimes we may use our own methods, such as click handlers. This is the beauty of a declarative UI toolkit: we don't even need to know how the UI we compose is actually rendered. We just use the API. However, in order to understand the intricacies of the Flutter build system, let's think about it for a moment.

How many times have you used `SizedBox` to add some spacing between other widgets? An interesting thing about this widget is that it extends neither `StatelessWidget` nor `StatefulWidget`. It extends `RenderObjectWidget`. As a developer, you will rarely need to extend this widget or any other that contains `RenderObjectWidget` in its title. The important thing to know about this widget is that it is responsible for rendering, as the name suggests. Each child of `RenderObjectWidget` has an associated `RenderObject` field. The `RenderObject` class is one of the three pillars of the Flutter build system (the first being the widget and the last being the `Element`, which we will see in the next section). This is the class that deals with actual low-level rendering details, such as translating user intentions onto the canvas.

Let's take a look at another example: the `Text` widget. Here is a piece of code for a very simple Flutter app that renders the `Hello, Flutter` text on the screen:

```
void main() {
  runApp(
    const MaterialApp(
      home: Text('Hello, Flutter'),
    ),
  );
}
```

We use two widgets here: the `MaterialApp`, which extends `StatefulWidget`, and `Text`, which extends a `StatelessWidget`. However, if we take a deeper look inside the `Text` widget, we will see that from its `build` method, a `RichText` widget is returned:

```
// Some code has been omitted for demo purposes
class Text extends StatelessWidget {
  @override
  Widget build(BuildContext context) {
    return RichText(...);
  }
}
class RichText extends MultiChildRenderObjectWidget {...}
```

An important difference here is that `RichText` extends `MultiChildRenderObjectWidget`, which is just a subtype of a `RenderObjectWidget`. So even though we didn't do it explicitly, the last widget in our widget tree extends `RenderObjectWidget`. We can visualize the widget tree, and it will look something like this:

The Widget Tree

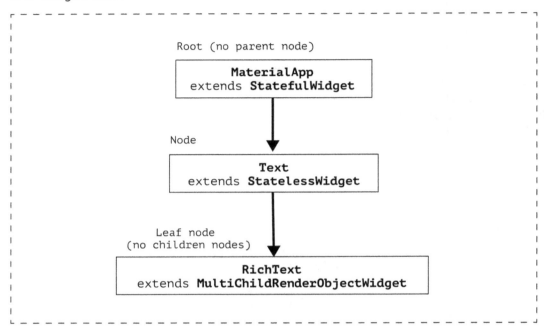

Figure 1.1 – Visual example of a widget tree

Even though you, as a developer, won't be extending `RenderObjectWidget` often, you need to remember one takeaway!

This is important!

No matter how aggressively you compose your widget tree, the widgets that are actually responsible for the rendering will always extend the `RenderObjectWidget` class. Even if you, as a developer, don't do it explicitly, you should know that this is what is happening deeper in the widget tree. You can always verify this by following the nesting of the `build` methods.

Let's sum up what we've learned about the widget types:

	StatelessWidget and StatefulWidget	**RenderObjectWidget**
Function	Composing widgets	Rendering render objects
Methods	`build`	`createRenderObject` `updateRenderObject`
Extended by developer	Often	Rarely
Examples	`Container`, `Text`	`SizedBox`, `Column`

Table 1.1 – Widget differences

But if widgets are immutable, then who updates the render objects?

Unveiling the Element class

From the `createRenderObject` and `updateRenderObject` we understand that render objects are mutable. Yet the widgets themselves that create those render objects are immutable. So how can they update anything, if they are recreated every time their `build` method is called?

The secret lies within the `Widget` API itself. Let's take a closer look at some of its methods, starting with `createElement`:

```
@immutable
abstract class Widget {

  Element createElement();

}
```

The first method that should interest us is createElement, which returns an Element. The element is the last of the three pillars of the Flutter build system. It does all of the shadow work, giving the spotlight to the widget. createElement gets called the first time the widget is added to the widget tree. The method calls the constructor of the overriding Element, such as StatelessElement. Let's take a look at what happens in the constructor of the Element class:

```
abstract class Element {

  Widget? _widget;

  Element(Widget widget)
      :_widget = widget {...}

}
```

We pass the widget field as the parameter to the constructor and assign it to the local _widget field. This way, the Element retains the pointer to the underlying widget, yet the widget doesn't retain the pointer to the element. The _widget field of the element is not final, which means that it can be reassigned. But when? The framework calls the update method of the Element any time the parent wishes to change the underlying widget. Let's take a look inside the update method source code:

```
abstract class Element {
  void update(covariant Widget newWidget) {
    _widget = newWidget;
  }
}
```

As we can see, the pointer to the _widget field is changed to newWidget, so the old widget is thrown away, yet our element stays the same. But in order for this reassignment to happen and for this method to be called, firstly the canUpdate method of the Widget class is called. The canUpdate method checks whether the runtimeType and key of the old and new widgets are the same as follows:

```
abstract class Widget {
  static bool canUpdate(Widget oldWidget, Widget newWidget) {
      return oldWidget.runtimeType == newWidget.runtimeType
          && oldWidget.key == newWidget.key;
  }
}
```

Only if this method returns true, which means that the runtimeType and key of the old and new widgets are the same, can we update our element with a new widget. Otherwise, the whole subtree will be disregarded and a completely new element will be inserted in this place.

To better understand the flow of this process, let's take a look at the following diagram:

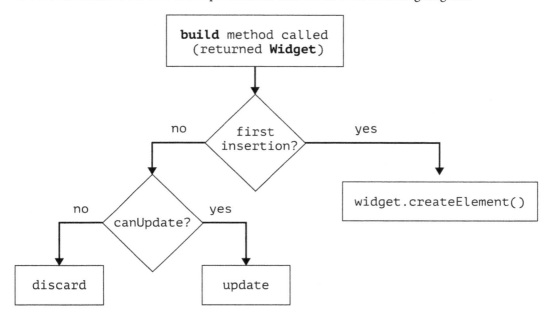

Figure 1.2 – Element relationship with the widget

The fact that the element can be updated instead of being recreated is even more important for the performance of RenderObjectWidget, since it deals with render objects that do the low-level painting. In the update method of RenderObjectElement, we also call updateRenderObject, which is a performance-optimized method: it only updates the render objects if there are any changes, and it only updates them partially. That's why even though there may be many calls of the build method, it doesn't mean that the whole tree gets completely repainted.

Finally, let's summarize everything we've learned about the Flutter tree system:

	Widget	Element	RenderObject
Mutable?	No	Yes	Yes
Cheap to create?	Yes	No	No
Created on every build?	Yes	No	No
Used by devs	Always	Almost never	Very rarely
Relationships	Every widget has an `Element`, but not every widget has a `RenderObject`	Implements `BuildContext` and has access both to the widget, and the `RenderObject` (if it exists)	Only created by implementers of `RenderObjectWidget`

Table 1.2 – Summary of widget, Element, and RenderObject roles

As we have just seen, Flutter has some straightforward yet elegant algorithms that make sure your application runs smoothly and looks flawless. Unfortunately, it doesn't mean that the developer doesn't have to think about performance at all, since there are many ways the performance can be impacted negatively if the best practices are ignored. Let's take a look at how we can support Flutter in maintaining a delightful experience for our users.

Reduce, reuse, recycle!

Now that we know that the `build` method can be called up to 120 times per second, the question is: do we really need to call the whole `build` method of our app if only a small part of the widget tree has changed? The answer is no, of course not. So let's review how we can make this happen.

First things first, let's get one obvious but still important thing out of the way. The `build` method is supposed to be blazing fast. After all, it can have as little as 8 ms to run without dropping frames. This is why it's crucial to keep any long-running tasks such as network or database requests out of this method. There are better places to do that which we will explore in detail throughout this book.

Pushing rebuilds down the tree

There can be several situations when pushing the rebuilds down the tree can impact performance in a positive way.

Calling setState of StatefulWidget

One of the most used widgets is `StatefulWidget`. It's a very convenient type of widget because it can manage state changes and react to user interactions. Let's take a look at the sample app that is created every time you start a new Flutter project: the counter app. We are interested in the code of the `_MyHomePageState` class, which is the `State` of `MyHomePage`:

```
class _MyHomePageState extends State<MyHomePage> {
  int _counter = 0;

  void _incrementCounter() {
    setState(() { _counter++; });
  }

  @override
  Widget build(BuildContext context) {
    return Scaffold(
      appBar: AppBar(
        title: const Text('Flutter Demo Home Page'),
      ),
      body: Center(
        child: Column(
          mainAxisAlignment: MainAxisAlignment.center,
          children: <Widget>[
            const Text(
              'You have pushed the button this many times:',
            ),
            Text(
              '$_counter',
              style: Theme.of(context).textTheme.headlineMedium,
            ),
            // In the original Flutter code the increment button is a
            // FloatingActionButton property of the Scaffold,
            // but for demonstration purposes, we need a slightly
            // modified version
            TextButton(
              onPressed: _incrementCounter,
              child: const Text('Increase'),
            )
          ],
        ),
      ),
    );
```

```
    }
  }
```

The UI is very simple. It consists of a `Scaffold` with an `AppBar` and a `FloatingActionButton`. Clicking the `FloatingActionButton` increments the internal `_counter` field. The body of the `Scaffold` is a `Column` with two `Text` widgets that describe how many times the `FloatingActionButton` has been clicked based on the `_counter` field. The preceding example differs from the original Flutter sample in one regard: instead of using the `FloatingActionButton` for handling clicks, we are using the `TextButton`. So every time we click the `TextButton`, the `_incrementCounter` method is called, which in turn calls the `setState` framework method and increments the `_counter` field. Under the hood, the `setState` method causes Flutter to call the `build` method of `_MyHomePageState`, which causes a rebuild. An important thing here is that `setState` causes a rebuild of the whole `MyHomePage` widget, even though we are only changing the text.

An easy way to optimize this is to push state changes down the tree by extracting them into a smaller widget. For example, we can extract everything that was inside the `Center` widget of `Scaffold` into a separate widget and call it `CounterText`:

```
class _MyHomePageState extends State<MyHomePage> {
  @override
  Widget build(BuildContext context) {
    return Scaffold(
      appBar: AppBar(
        title: const Text('Flutter Demo Home Page'),
      ),
      body: const Center(child: CounterText()),
    );
  }
}

class CounterText extends StatefulWidget {
  const CounterText({Key? key}) : super(key: key);

  @override
  State<CounterText> createState() => _CounterTextState();
}

class _CounterTextState extends State<CounterText> {
  int _counter = 0;

  void _incrementCounter() {
    setState(() { _counter++; });
```

```
  }

  @override
  Widget build(BuildContext context) {
    return Column(...// Same code that was in the original example );
  }
}
```

We haven't changed any logic. We only took the code that was inside of the Center widget of _
MyHomePageState and extracted it into a separate widget:CounterText. By encapsulating the
widgets that need to be rebuilt when an internal field changes into a separate widget, we ensure that
whenever we call setState inside of the _CounterTextState field, only the widgets returned
from the build method of _CounterTextState get rebuilt. The parent _MyHomePageState
doesn't get rebuilt, because its build method wasn't called. We pushed the state changes down the
widget tree, causing only smaller parts of the tree to get rebuilt, instead of the whole screen. In real-
life app code, this scales very fast, especially if your pages are UI-heavy.

Subscribing to InheritedWidget changes via .of(context)

By extracting the changing counter text into a separate CounterText widget in the last code
snippet, we have actually made one more optimization. The interesting line for us is Theme.
of(context).textTheme.headlineMedium. You have certainly used Theme and other
widgets, such as MediaQuery or Navigator, via the .of(context) pattern. Usually, those
widgets extend a special type of class: InheritedWidget. We will look deeper into its internals in
the state management part (*Chapters 3* and *4*), but for now, we are interested in two of its properties:

- Instead of creating those widgets, we will access them via static getter and use some of their
  properties. This means that they were created somewhere higher up the tree. Hence, we will
  inherit them. If they weren't and we still try to look them up, we will get an error.

- For some of those widgets, such as Theme and MediaQuery, the .of(context) not only
  returns the instance of the widget if it finds one but also adds the calling widget to a set of its
  subscribers. When anything in this widget changes – for example, if the Theme was light and
  became dark – it will notify all of its subscribers and cause them to rebuild. So in the same way
  as with setState, if you subscribe to an InheritedWidget, changes high up in the tree
  will cause the rebuild of the whole widget tree starting from the widget that you have subscribed
  in. Push the subscription down to only those widgets that actually need it.

Extra performance tip

You may have used `MediaQuery.of(context)` in order to fetch information about the screen, such as its size, paddings, and view insets. Whenever you call `MediaQuery.of(context)`, you subscribe to the whole `MediaQuery` widget. If you want to get updates only about the paddings (or the size, or the view insets), you can subscribe to this *specific property* by calling `MediaQuery.paddingOf(context)`, `MediaQuery.sizeOf(context)`, and so on. This is because `MediaQuery` actually extends a specific type of `InheritedWidget` – the `InheritedModel` widget. It allows you to subscribe only to those properties that you care about as opposed to the whole widget, which can greatly contribute to widget rebuild optimization.

Avoiding redundant rebuilds

Now that we've learned how to scope our trees so that only smaller sections are rebuilt, let's find out how to minimize the amount of those rebuilds altogether.

Being mindful of the widget life cycle

Stateless widgets are boring in terms of their life cycles. Stateful widgets, on the other hand, are not. Let's take a look at the life cycle of the `State`:

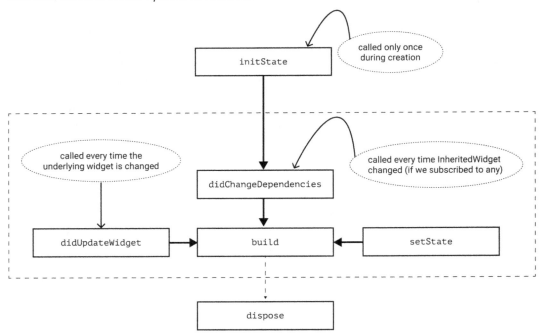

Figure 1.3 – Main methods of State life cycle

Here are a few things that we should care about:

- The `initState` method gets called only once per widget life cycle, much like the `dispose` method.

- The `didChangeDependencies` method gets called immediately after `initState`.

- `didChangeDependencies` is always called when an `InheritedWiget` that we subscribed to has changed. This is the implementation aspect of what we have just discussed in the previous section.

- The build method always gets called after `didChangeDependencies`, `didUpdateWidget`, and `setState`.

> **This is important!**
> Don't call `setState` in `didChangeDependencies` or `didUpdateWidget`. Such calls are redundant, since the framework will always call `build` after those methods.

The best performance practices in the preceding list are also the reason why it's better to decouple your widgets into other custom widgets rather than extract them into helper methods such as `Widget buildMyWidget()`. The widgets extracted into methods still access the same context or call `setState`, which causes the whole encapsulating widget to rebuild, so it's generally recommended to prefer widget classes rather than methods.

One more important thing regarding the life cycle of the `State` is that once its `dispose` method has been called, it will never become alive again and we will never be able to use it again. This means that if we have acquired any resources that hold a reference to this `State`, such as text editing controllers, listeners, or stream subscriptions, these should be released. Otherwise, the references to these resources won't let the garbage collector clean up this object, which will lead to memory leaks. Fortunately, it's usually very easy to release resources by calling their own `dispose` or `close` methods inside the `dispose` of the `State`.

Caching widgets implicitly

Dart has a notion of constant constructors. We can create constant instances of classes by adding a `const` keyword before the class name. But when can we do this and how can we take advantage of them in Flutter?

First of all, in order to be able to declare a `const` constructor, all of the fields of the class must be marked as `final` and be known at compile time. Second, it means that if we create two objects via `const` constructors with the same params, such as `const SizedBox(height: 16)`, only one instance will be created. Aside from saving some memory due to initializing fewer objects, this also provides benefits when used in a Flutter widget tree. Let's return to our `Element` class once again.

We remember that the class has an `update` method that gets called by the framework when the underlying widget has changed its fields (but not type or key). This method changes the reference to the widget. Soon the framework calls rebuild. Since we're working with a tree data structure, we will traverse its children. Unless your element is a leaf element, it will have children. There is a very important method in the `Element` API called `updateChild`. As the name says, it updates its children elements. But the interesting thing is how it does it:

```
#1 Element? updateChild(Element? child, Widget? newWidget,
    Object? newSlot) {
#2    // A lot of code removed for demo purposes
#3
#4    final Element newChild;
#5    if (child.widget == newWidget) {
#6        newChild = child;
#7    } else if (Widget.canUpdate(child.widget, newWidget)) {
#8        child.update(newWidget);
#9        newChild = child;
#10   }
#11
#12   return newChild;
#13 }
```

In the preceding code, in case our current widget is the same as the new widget as determined by the `==` operator, we only reassign the pointer, and that's it. By default, in Dart, the `==` operator returns `true` only if both of the instances point to the same address in memory, which is `true` if they were created via a `const` constructor with the same params.

However, if the result is `false`, we should check the already-familiar `Widget.canUpdate`. However, aside from reassigning the pointer to the new element, we also call its `update` method, which soon causes a rebuild.

Hence, if we use `const` constructors, we can avoid rebuilds of whole widget subtrees. This is also sometimes referred to as caching widgets. So use `const` constructors whenever possible and see whether you can extract your own widgets that can make use of `const` constructors, even if nested widgets can't.

Keep in mind that you have to actually use the `const` constructor, not just declare it as a possibility. For example, we have a `ConstText` widget that has a `const` constructor:

```
class ConstText extends StatelessWidget {
  const ConstText({super.key});

  @override
  Widget build(BuildContext context) {
    return const Text('Hello World');
```

```
  }
}
```

However, if we create an instance of this widget without using the const constructor via the const keyword as in the following code, then we won't get any of the benefits of the const constructor:

```
// Don't!
class ParentWidget extends StatelessWidget {
  const ParentWidget({Key? key}) : super(key: key);

  @override
  Widget build(BuildContext context) {
    return ConstText(); // not const!
  }
}
```

We need to explicitly specify the const keyword when creating an instance of the class. The correct usage of the const constructor looks like this:

```
// Do
class ParentWidget extends StatelessWidget {
  const ParentWidget({Key? key}) : super(key: key);

  @override
  Widget build(BuildContext context) {
    return const ConstText(); // const, all good
  }
}
```

In the preceding code, we used the const keyword during the creation of a ConstText widget. This way, we will get all of the benefits. This small keyword is very important.

Explicitly cache widgets

The same logic can be applied if the widget can't be created with a const constructor, but can be assigned to a final field of the State. Since you're literally saving the pointer to the same widget instance and returning it rather than creating a new one, it will follow the same execution path as the one we saw with const widgets. This is one of the ways in which you can work around the Container not being const. You might do so using the following, for example:

```
class _MyHomePageState extends State<MyHomePage> {
  final greenContainer = Container(
    color: Colors.green,
    height: 100,
    width: 100,
```

```
    );

    @override
    Widget build(BuildContext context) {
      return Column(
        children: [
          greenContainer,
          Container(
            color: Colors.pink,
            height: 100,
            width: 100,
          ),
        ],
      );
    }
}
```

In the preceding code, the `update` method of the green container won't be called. We have retained the reference to an already-existing widget by caching it in a local `greenContainer` field. Hence, we return the exact same instance as in the previous `build`. This falls into the case described on line 5 in the `updateChild` method code snippet provided earlier in this section. If the instances are the same based on the equality operator, then the `update` method is not called. On the other hand, the pink `Container` will be rebuilt every time because we create a new instance of the class every time the `build` method is called. This is described in line 7 of the same code snippet.

Avoiding redundant repaints

Up to this point, we have looked at tips to help you avoid causing redundant rebuilds of the widget and element trees. The truth is that the building phase is quite cheap when compared to the rendering process, as this is where all of the heavy lifting is done. Flutter optimizes this phase as much as possible, but it cannot completely control how we create our interfaces. Therefore, we may encounter cases where these optimizations are not enough or are not working effectively.

Let's take a look at what happens when one of the render objects wants to repaint itself. We may assume that this repainting is scoped to that specific object – after all, it was the only one marked for repaint. But this is not what happens.

The thing is, even though we have scoped our widget tree, our render object tree has a relationship of its own. If a render object has been marked as needing repainting, it will not only repaint itself but also ask its parent to mark itself for repaint too. That parent then asks its parent, and so on, until the very root. And when it finally comes to painting, the object will also repaint all of its descendants. This happens until the framework encounters what is known as a **repaint boundary**. A repaint boundary is a Flutter way of saying "stop right here, there is nothing further to repaint." This is done by wrapping your widget into another widget – yes, the `RepaintBoundary` widget.

If we wanted to depict this flow visually, it would be something like this:

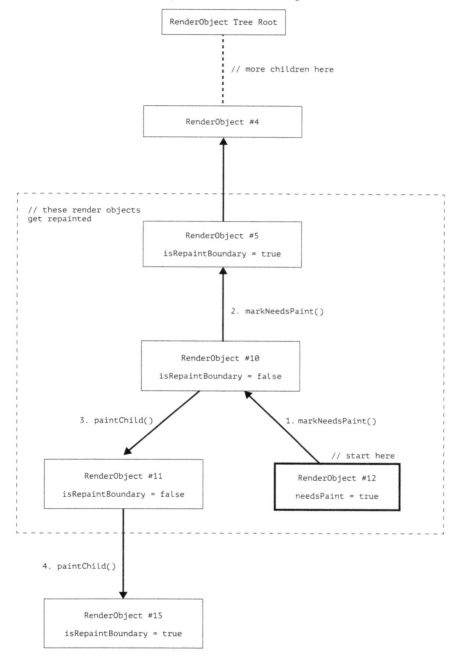

Figure 1.4 – Flow of the render object's repainting process

Here is what's happening in *Figure 1.4*:

1. We start from **RenderObject #12**, which was the initial one to be marked for repainting.
2. The object goes on to call **parent.markNeedsPaint** of **RenderObject #10**. Since the **isRepaintBoundary** field is **false**, the **needsPaint** gets set to **true** and goes on to ask the same for its parent.
3. The **isRepaintBoundary** value of **RenderObject #5** is **true**, so **needsPaint** stays **false** and the parent marking is stopped right there.
4. Then the actual painting phase is started from the top widget marked as **needsPaint**. It traverses its children. Since **isRepaintBoundary** of **RenderObject #11** is **false**, it traverses further.
5. But **isRepaintBoundary** of **RenderObject #15** is **true**, so the process is stopped right there.

So we end up repainting render objects #10, #11, and #12.

Let's take a look at an example where a `RepaintBoundary` widget can be useful – in a `ListView`. This is the simplified version of the `ListView` source code:

```
class ListView {

  ListView({
      super.key,
      bool addRepaintBoundaries = true,
      ... // many more params
  });

  @override
  Widget? build(BuildContext context, int index) {
      if (addRepaintBoundaries) {
        child = RepaintBoundary(child: child);
      }
      return child;
  }
}
```

The `ListView` constructor accepts an `addRepaintBoundaries` parameter in its constructor, which by default is `true`. Later, when building its children, the `ListView` checks this flag, and if it's `true`, the child widget is wrapped in a `RepaintBoundary` widget. This means that during scrolling, the list items don't get repainted, which makes sense because only their offset changes, not their presentation. The `RepaintBoundary` widget can be extremely efficient in cases where you have a heavy yet static widget, or when only the location on the screen changes such as during scrolling, transitions, or other animations. However, like many things, it has trade-offs. In order to display the end result on the screen, the widget tree drawing instructions need to be translated into the actual pixel data. This process is called **rasterization**. `RepaintBoundary` can decide to cache the rasterized pixel values in memory, which is not limitless. Too many of them can ironically lead to performance issues.

There is also a good way to determine whether the `RepaintBoundary` is useful in your case. Check the `diagnosis` field of its `renderObject` via the Flutter inspector tools. If it says something along the lines of **This is an outstandingly useful repaint boundary**, then it's probably a good idea to keep it.

Optimizing scroll view performance

There are two important tips for optimizing scroll view performance:

- First, if you want to build a list of homogeneous items, the most efficient way to do so is by using the `ListView.builder` constructor. The beauty of this approach is that at any given time, by using the `itemBuilder` callback that you've specified, the `ListView` will render only those items that can actually be seen on the screen (and a tiny bit more, as determined by the `cacheExtent`). This means that if you have 1,000 items in your data list, you don't need to worry about all 1,000 of them being rendered on the screen at once – unless you have set the `shrinkWrap` property to `true`.

- This leads us to the second tip: the `shrinkWrap` property (available for various scroll views) forces the scroll view to calculate the layout of all its children, defeating the purpose of lazy loading. It's often used as a quick fix for overflow errors, but there are usually better ways to address those errors without compromising performance. We'll cover how to avoid overflow errors while maintaining performance in the next chapter.

Summary

In this chapter, we explored the relationships between the `Widget`, `Element`, and `RenderObject` trees. We learned how to avoid rebuilds of the `Widget` and `Element` trees by scoping the `StatefulWidgets` and subscriptions to inherited widgets, as well as by caching the widgets via `const` constructors and `final` initializations. We also learned how to limit repaints of the render object subtrees, as well as how to effectively work with scroll views.

In *Chapter 2*, we will explore how to make our already performant interfaces responsive on the ever-growing set of devices. We will cover how sizing and layout work in Flutter, how to fix overflow errors, and how to ensure that your application is usable for all users.

Get this book's PDF version and more

Scan the QR code (or go to `packtpub.com/unlock`). Search for this book by name, confirm the edition, and then follow the steps on the page.

UNLOCK NOW

Note: Keep your invoice handy. Purchases made directly from Packt don't require an invoice.

2
Responsive UIs for All Devices

Flutter originally started as a framework with a primary focus on creating high-quality native apps for iOS and Android smartphones. It quickly evolved to support platforms beyond smartphones, such as tablets, web browsers, and desktop applications. With such a variety of devices to support, it has become important for developers to learn how to create responsive **user interfaces** (**UIs**) that look great on every screen. The key to creating these responsive UIs is to use techniques and strategies that enable the UI to adapt to different screen sizes, orientations, and resolutions.

In this chapter, we will get to explore how Flutter lays out the different widgets on screen and how this integrates with the build process that we covered in the first chapter. Secondly, we will learn how to obtain information about the device's screen size and orientation and use it to adjust the UI based on these parameters. We will also discuss which widgets can help us control the position and size of our widgets and allow us to create complex and flexible layouts. Finally, we will cover how to implement **accessibility** features seamlessly into our Flutter layouts, ensuring that our beautifully designed interfaces are usable and welcoming to individuals of all abilities.

By the end of this chapter, you will have a better understanding of how the layout phase works in Flutter and how to create responsive and adaptive UIs using Flutter's layout system. As Flutter continues to evolve and support more platforms, the importance of creating responsive UIs will only continue to grow, making it essential for developers to master these strategies.

In this chapter, we're going to cover the following main topics:

- Understanding the Flutter layout algorithm
- Designing responsive apps with Flutter
- Ensuring accessibility in Flutter apps

Technical requirements

If this is not your first Flutter app, then you probably already have everything you need installed.

Otherwise, you will need to install the following:

- An IDE of your choice that supports Flutter, such as Android Studio or VSCode
- The Flutter SDK

All of the code samples referred in this chapter can be found here: `https://github.com/PacktPublishing/Flutter-Design-Patterns-and-Best-Practices/tree/master/CH02/candy_store`.

Understanding the Flutter layout algorithm

In Flutter, the layout phase is an essential part of the widget-building process. It is during this phase that the engine calculates how the widgets are positioned and sized on the screen. The layout phase starts after the build process that we discussed in *Chapter 1* is done and is followed by the painting and composition steps of the rendering pipeline.

This layout process is critical to the overall performance and user experience of the app. If the layout is incorrect, widgets may overlap or be positioned in unexpected locations on the screen, leading to a confusing and frustrating experience for the user.

In Flutter, the layout process is handled by the `RenderObject` class. Each `RenderObjectWidget` in the widget tree has a corresponding `RenderObject` that is responsible for handling the layout and rendering of that widget. The `RenderObject` calculates its size based on information provided by its parent `RenderObject` and the position of its child render objects. This information passed by the parent to its child is called **constraints**. Let's learn why constraints are so important.

Understanding BoxConstraints

To understand the layout process, we first need to understand the concept of constraints. In Flutter, each widget has a set of constraints that define its minimum and maximum allowable size. While the abstract version of `Constraints` is not bound to anything and allows the implementer to be as creative as needed, it will be easier to understand constraints if we use a more concrete example. The `BoxConstraints` method is the implementation that describes the constraints based on the notion of a box that can have a minimum and maximum width and height. It is used for many, if not most, of the widgets that you will encounter. There are multiple possibilities for such constraints. For example, a widget could have a minimum 0 size and a maximum infinite size. To set such constraints, we would specify them in a BoxConstraints constructor like this:

```
BoxConstraints(
  minWidth: 0,
  minHeight: 0,
  maxWidth: double.infinity,
  maxHeight: double.infinity,
);
```

These constraints would mean that the sizing options are endless and that our widget is free to choose the size it wants. On the other hand, we could be more strict and force the widget to match only a specific size, with a width of 10 and a height of 20. The BoxConstraints need to be applied to a widget that accepts constraints as a parameter, such as ConstrainedBox. We can specify a colorful child, such as a ColoredBox. Our code would look like this:

lib/example_constraints.dart

```
void main() {
  runApp(
    MaterialApp(
      home: Scaffold(
        body: ConstrainedBox(
          constraints: BoxConstraints(
            minWidth: 10,
            minHeight: 20,
            maxWidth: 10,
            maxHeight: 20,
          ),
          child: ColoredBox(color: Colors.red),
        ),
      ),
    ),
  );
}
```

In this case, the widget will be forced to have a width of 10 and height of 20, with no other option, since the minimum allowed size is the same as the maximum. The size of the widget will always need to satisfy its constraint conditions.

> **Fun fact**
>
> In the preceding snippet, we have done what Container does under the hood by wrapping different kinds of box widgets based on parameters such as constraints, size, and color.

How do constraints determine the child widget's size?

The Flutter layout algorithm can be summarized as follows:

Constraints go down. Sizes go up. Parent sets the position
`https://docs.flutter.dev/ui/layout/constraints`

This easy and catchy rule describes the three key steps that are needed to position and lay out any widget.

Let's break it down in more detail:

1. **Constraints go down**: A widget gets its own constraints from its parent. This widget also passes new constraints to its children one by one (these can be different for each child). This process walks down the entire render tree until it reaches the last child. Once every `RenderObject` has constraints, the next step is to define the actual size.

2. **Sizes go up**: The child picks a size that satisfies the constraints received by the parent. The sizing algorithm is defined by each `RenderObject` depending on their own needs. In this step, together with *step 3*, the framework walks back up the render tree passing the defined geometry.

3. **Parent sets the position**: Once all children have had their size defined, the parent will position its children (horizontally on the *x* axis and vertically on the *y* axis) one by one. Once all children have been positioned, the parent will move to *step 2* again until we reach the `RenderObject` root.

This can be visualized with the following diagram:

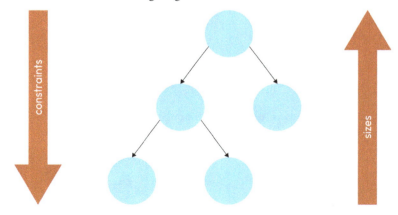

Figure 2.1 – Constraints going down from the parent to its children
and sizes going up from the children to the parent

In the diagram, you can see that the constraints are passed all the way down from the root to the leaf nodes of the render object tree. Then the sizes are passed all the way up from the leaf nodes to the root. This also demonstrates the efficiency of Flutter's layout algorithm: the layout process is a single-pass process. This means that the render tree is walked at most once in each direction: down to pass constraints and up to pass sizes. Now let's look at some specific examples.

The parent widget limits the size of the child with the passed constraints but it is the child that decides its final size.

Let's see this through a simple example of a container and a Text widget as its child:

Figure 2.2 – A red container with Hello World text aligned to the top-left corner

For brevity, we will only look at the code of the relevant widget. You can find the fully runnable example in the filename mentioned at the start of the code snippet. Until you see otherwise otherwise, keep in mind that the Container that we're working with is a child of a Scaffold.

The code to generate this widget is as follows:

```
Container(
    constraints: BoxConstraints(
      minWidth: 200,
      minHeight: 100,
      maxWidth: 200,
      maxHeight: 100,
    ),
    color: Colors.red,
    child: Text('Hello World'),
)
```

In the preceding code, the Container allows any child that can fit inside it, as long as the child satisfies the constraints. As a result, the RenderObject used by Text will take these constraints into consideration when calculating its geometry.

Something interesting that you may have noticed in this code is that we set the same values for minWidth and maxWidth, as well as for minHeight and maxHeight. In terminology, this is known as **tight constraints** because there is quite literally no wiggle room. The size is tightly constrained. For this, BoxConstraints has a BoxConstraints.tight constructor. Instead of accepting four values, this requires just two. Code like this will produce the same result as the previous snippet:

```
Container(
  constraints: BoxConstraints.tight(
    const Size(200, 100),
```

```
    ),
    color: Colors.red,
    child: Text('Hello World'),
)
```

On the other hand, if we want to specify only the maximum size and let the minimum size be 0 so that the child widget can decide for itself, this is called **loose constraints**. In a similar way, to avoid specifying all four values, we can use the `BoxConstraints.loose` constructor like this:

```
Container(
  constraints: BoxConstraints.loose(
    const Size(200, 100),
  ),
  color: Colors.red,
  child: Text('Hello World'),
)
```

The result will be exactly the same if we've written it like this:

```
Container(
  constraints: BoxConstraints(
    minWidth: 0,
    minHeight: 0,
    maxWidth: 200,
    maxHeight: 100,
  ),
  color: Colors.red,
  child: Text('Hello World'),
)
```

With loose constraints, our text would only take up as much space as it requires. Hence, the container will do the same. Instead of being the `100x200` size, it will wrap the text tightly like this:

Hello World

Figure 2.3 – A red container with Hello World text wrapped with loose constraints

Once the child knows its size, the parent will take care of positioning it. The parent defines how its children will be allocated on the screen. It is important to remark that a child does not know its own position. In the previous example, the container decided to position the `Text` widget child in the top left of the screen, while the child only knows its size.

Now we can modify the position of the child by adding or modifying its parents. Imagine that we want to center the text inside the container in the previous example, as shown in the following figure.

Figure 2.4 – A red container with Hello World text aligned to the center

As the container is the one that defines the position, we will have to either modify it or wrap the text with a more appropriate parent that has a different positioning algorithm. In this case, we will change the alignment field of this widget as follows:

lib/example_container_5.dart

```
Container(
  alignment: Alignment.center,
  constraints: BoxConstraints.tight(Size(200, 100)),
  color: Colors.red,
  child: Text('Hello World'),
);
```

Container is a very flexible widget and presents us with many customization options. In the preceding code, we added a property called alignment with the Alignment.center value. There are many more options such as Alignment.topRight or Alignment.bottomLeft. However, not all widgets are as flexible as Container, or maybe you can't use the alignment property for another reason. In that case, you can wrap your child widget, as we did with our Text, in a special widget called Align. Its sole purpose is to align the child widget. Having multiple options for various use cases is great and Flutter gives us just that. Here's how we could use the Align widget to achieve the same effect:

```
Container(
  constraints: BoxConstraints.tight(
    const Size(200, 100),
  ),
  color: Colors.red,
  child: Align(
    alignment: Alignment.center,
    child: Text('Hello World'),
  ),
),
```

Aside from the `Align` widget, there is also a very specific one called `Center`, which aligns the child widget in the center and provides the developer with a more concise API.

> **Note**
>
> There are endless scenarios that can occur when defining constraints and positioning children. To see more examples, you can check out `https://docs.flutter.dev/development/ui/layout/constraints`.

Understanding the limitations of the layout rule

While the layout algorithm has quite a lot of benefits, such as only needing one pass through the whole tree, it also comes with some limitations. It is important to be aware of the limitations so that we don't try to fight against them. Let's see what some of them are:

- **A widget might not have its desired size**: A child is forced to follow its parent's constraints, yet the child's desired size might not fit in those constraints, so the child's final size might not match the desired one. One scenario where this can create conflicts is when the parent forces its child to fit a tight size. In our previous experiments with the `Container` widget, it has always been a child of the `Scaffold`. The `Scaffold` imposes its own constraints, which is why the results we have seen have corresponded with our expectations. However, there are use cases that might surprise you. In the following code, we set a container as the root widget, with a desired size of `10x10`:

lib/example_container_7.dart

```
runApp(
  Container(
    height: 10,
    width: 10,
    color: Colors.red,
  ),
);
```

While we would expect the `Container` to have a `10x10` size, this won't be the actual case and the `Container` will cover the full screen. It will look like this:

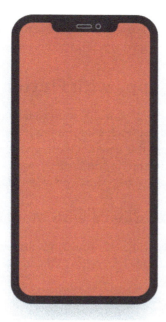

Figure 2.5 – A red container filling the full screen

The `height` and `width` params in the preceding code are in reality a shortcut to define the constraints of its child. They are defined with`BoxConstraints.tightFor(width: width, height: height)`.

So, let's see what is happening in the preceding code based on what we have learned so far. First, the `Container` receives constraints from its parent. As this `Container` is the top widget, it is a special case. The `RenderObject` root will have a tight constraint that *enforces the minimum and maximum to be the size of the screen*. Therefore, when the `Container` is sizing itself, the only option available is the size of the screen.

- **A widget is not aware of its own position on the screen**: Another interesting limitation is that we cannot learn the position of a widget from the widget itself. Its parent contains that information. While this looks like a huge disadvantage, it has a huge performance benefit for algorithms that do not have to recalculate child size each time they want to position their items (for example, `GridView`). In the upcoming sections of this chapter, we will learn how we can create widgets that can be positioned relative to others and therefore mitigate this inconvenience.

- **Final size and position are relative to a parent**: The final consideration is that a widget's size and position depend on its parent. This parent widget also depends on its own parent. It is impossible to define the size and position of a specific widget without taking the whole tree into consideration.

Now that we know the general rule and its limitations, let's learn about the layout solutions that implement these basic concepts. We will learn how we can use them to create our own layouts. When in doubt, always remember: constraints go down, sizes go up, parent sets the position.

Designing responsive apps with Flutter

When building your Flutter app, you will probably want to target multiple platforms and screen sizes. It is important to keep in mind that each platform has its own unique design guidelines and user experience expectations. However, this does not mean giving up on designing a flexible layout that can adapt to most scenarios. Flutter brings with it all the tools needed to create responsive apps that work across a wide range of devices, screen sizes, and orientations. Let's discover what these are.

Getting to know the user's device with MediaQuery

In Flutter, MediaQuery is a utility widget that provides information about the device and its constraints. It allows you to query various properties of the device, such as its orientation, screen size, pixel density, and more.

> **Good to know**
>
> When talking about screen size and pixels, there are a couple of concepts that we need to understand. **Physical pixels**, also known as device pixels, is a measure of the number of actual pixels on the display. This can vary a lot based not only on the actual screen size but also on pixel density. Pixel density is often measured in inches and referred to as **pixel per inch (PPI)**. The higher the PPI, the higher the resolution and the better the display quality. On the other hand, there are **logical pixels** which are roughly 96 physical pixels per inch. This concept is also known as **device-independent pixels**. It allows for creating an app UI without dealing with pixel density. In Flutter, we work with logical pixels.

Here are some of the most useful properties we can use:

- size: The physical size of the device's screen in logical pixels
- devicePixelRatio: The number of physical pixels per logical pixel
- orientation: The orientation of the device (portrait or landscape)
- padding: The amount of padding applied to the screen, such as the status bar and navigation bar
- textScaleFactor: The scaling factor applied to text based on the device's accessibility settings

The MediaQuery widget is included in the widget tree when using MaterialApp, CupertinoApp, or WidgetsApp in your Flutter app. This allows you to access the properties of the device and its constraints throughout your app. For example, you might use MediaQuery.sizeOf(context) to retrieve the screen size of the device and adjust the layout of your app accordingly.

Here's an example of how you can use `MediaQuery` to create a layout that adapts to different screen sizes:

lib/example_responsive_layout.dart

```dart
void main() {
  runApp(
    MaterialApp(
      home: Scaffold(
        body: ResponsiveLayoutExample(),
      ),
    ),
  );
}

class ResponsiveLayoutExample extends StatelessWidget {
  const ResponsiveLayoutExample({super.key});

  @override
  Widget build(BuildContext context) {
    final screenSize = MediaQuery.of(context).size;

    return GridView.count(
      crossAxisCount: _getCrossAxisCount(screenSize.width),
      children: List.generate(
        12,
        (index) => Container(
          margin: const EdgeInsets.all(8),
          alignment: Alignment.center,
          color: Colors.green,
          child: Text('Item ${index + 1}'),
        ),
      ),
    );
  }

  int _getCrossAxisCount(double screenWidth) {
    if (screenWidth >= 1200) {
      return 4;
    } else if (screenWidth >= 800) {
      return 3;
    } else if (screenWidth >= 600) {
      return 2;
    } else {
```

```
        return 1;
      }
    }
  }
```

In the preceding example, we used `MediaQuery` to retrieve the screen size and determine the number of columns in the grid based on the screen width. The `_getCrossAxisCount` method returns 4 columns for screens wider than `1200` pixels, 3 columns for screens between `800` and `1200` pixels, 2 columns for screens between `600` and `800` pixels, and 1 column for screens narrower than `600` pixels. We have also used a `GridView` widget here, which is an easy way to show a list of items in a grid format. Here is how it would look on different devices:

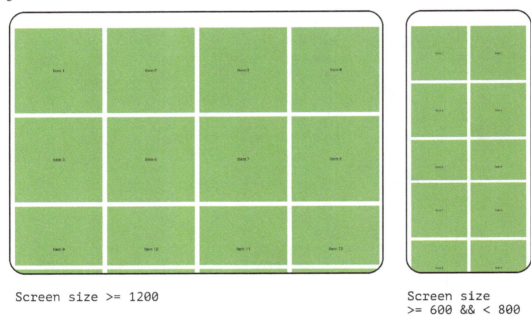

Screen size >= 1200

Screen size
>= 600 && < 800

Figure 2.6 – A responsive layout with various numbers of columns depending on the screen width

By using `MediaQuery` to adapt the layout to different screen sizes, we can ensure that our app looks good and is easy to use on a wide range of devices.

Creating adaptive layouts

When we want to create dynamic layouts based on the available size or the device's orientation, `LayoutBuilder` and `OrientationBuilder` are our safest bet.

LayoutBuilder is a widget that provides a callback that you can use to build your layout based on the available layout constraints. The widget allows you to create layouts that are responsive and adapt to different screen sizes. For example, you might use LayoutBuilder to adjust the number of columns in a grid layout based on the screen width or to adjust the font size of text based on the available height.

Here's an example of how to use LayoutBuilder:

lib/example_layout_builder.dart

```dart
void main() {
  runApp(
    MaterialApp(
      home: Scaffold(
        body: LayoutBuilder(
          builder: (BuildContext context, BoxConstraints constraints){
            if (constraints.maxWidth > 600) {
              return DesktopLayout();
            } else {
              return MobileLayout();
            }
          },
        ),
      ),
    ),
  );
}

class DesktopLayout extends StatelessWidget {
  const DesktopLayout({super.key});

  @override
  Widget build(BuildContext context) {
    return Container(
      color: Colors.pink,
      child: Align(
        alignment: Alignment.center,
        child: Text('Desktop Layout'),
      ),
    );
  }
}

class MobileLayout extends StatelessWidget {
```

```
    const MobileLayout({super.key});

    @override
    Widget build(BuildContext context) {
      return Container(
        color: Colors.purple,
        child: Align(
          alignment: Alignment.center,
          child: Text('Mobile Layout'),
        ),
      );
    }
  }
```

Here is how it would look on different devices:

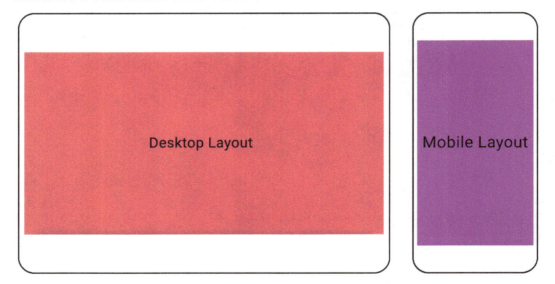

Figure 2.7 – A responsive layout with different layouts on mobile and desktop devices

In the preceding example, we used `LayoutBuilder` to switch between a desktop layout and a mobile layout based on the maximum width of the available layout constraints.

`OrientationBuilder`, on the other hand, is a widget that provides a callback that you can use to build your widgets based on the device's orientation. This allows you to create layouts that adjust to changes in orientation, such as rotating the device from portrait to landscape mode. For example, you might use `OrientationBuilder` to adjust the layout of your app when the device is rotated, such as by rearranging elements or adjusting the size and position of widgets.

Here's an example of how to use `OrientationBuilder`:

lib/example_orientation_builder.dart

```dart
void main() {
  runApp(
    MaterialApp(
      home: Scaffold(
        body: OrientationBuilder(
          builder: (BuildContext context, Orientation orientation) {
            if (orientation == Orientation.portrait) {
              return PortraitLayout();
            } else {
              return LandscapeLayout();
            }
          },
        ),
      ),
    ),
  );
}

class LandscapeLayout extends StatelessWidget {
  const LandscapeLayout({super.key});

  @override
  Widget build(BuildContext context) {
    return Container(
      color: Colors.green,
      child: Align(
        alignment: Alignment.center,
        child: Text('Landscape Layout'),
      ),
    );
  }
}

class PortraitLayout extends StatelessWidget {
  const PortraitLayout({super.key});

  @override
  Widget build(BuildContext context) {
    return Container(
```

```
            color: Colors.yellow,
            child: Align(
              alignment: Alignment.center,
              child: Text('Portrait Layout'),
            ),
          );
        }
    }
```

In the preceding example, we used `OrientationBuilder` to switch between a portrait layout and a landscape layout based on the device's orientation.

Overall, `LayoutBuilder` and `OrientationBuilder` are two useful utility widgets in Flutter that can help you create responsive and adaptable layouts that adjust to changes in screen size and orientation.

Positioning widgets relative to each other

`Stack` and `Align` are two of the most commonly used widgets in Flutter for arranging and positioning child widgets. We are already familiar with the `Align` widget, which allows us to change the alignment of the child widget. Now let's get to know the `Stack` widget.

The `Stack` widget allows you to stack widgets on top of each other, similar to a stack of cards. The child widgets are positioned relative to the top-left corner of the stack by default but you can also position them relative to other corners or to the center of the stack. The order in which the child widgets are added to the stack determines the order in which they are painted onto the screen, with the last child widget added appearing on top.

Here's an example of how to use the `Stack` widget:

lib/example_stack_1.dart

```
runApp(
  MaterialApp(
    home: Scaffold(
      body: Stack(
        children: [
          Container(
            width: 100,
            height: 100,
            color: Colors.red,
          ),
          Container(
            width: 50,
            height: 50,
```

```
      color: Colors.blue,
    ),
   ],
  ),
 ),
),
);
```

As a result of the preceding code, the two containers will be laid out as follows.

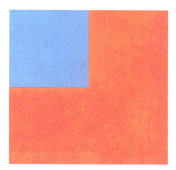

Figure 2.8 – A stack with two colored containers

The preceding code will display a blue square on top of a red square since the blue square was added after the red square.

To position widgets within a stack, the `Positioned` widget can be used. The `Positioned` widget can only be added as a child of the `Stack` widget. It provides a way to position its child widget in a specific location within the stack. Using `Positioned` when it's not a direct child of a `Stack` widget will throw an `Incorrect use of ParentDataWidget` error.

Here's an example of how to use the `Positioned` widget to position a blue box in the bottom-right corner of the stack. We will omit the code required to run the app, as it stays the same as in the previous example, and just focus on the `Stack`:

lib/example_stack_2.dart

```
Stack(
  children: [
    Container(
      width: 100,
      height: 100,
      color: Colors.red,
    ),
    Positioned(
```

```
          bottom: 0,
          right: 0,
          child: Container(
            width: 50,
            height: 50,
            color: Colors.blue,
          ),
        ),
      ],
    )
```

In this case, the two containers will be laid out with different alignments, as shown in the following figure.

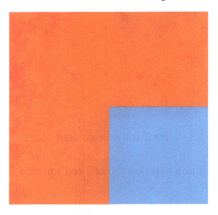

Figure 2.9 – A stack with two colored containers with different alignments

As you can see in the preceding code, we created a blue box using the Container widget, then we positioned it in the bottom-right corner of the stack using the Positioned widget.

The Align widget, on the other hand, allows you to position a child widget within another widget using a combination of horizontal and vertical alignment properties. The child widget is first positioned within the container according to the given horizontal and vertical alignment properties, then any additional offset properties are applied. By combining the Stack, Positioned, and Align widgets, you can create complex layouts with precise positioning and layering of child widgets. In case you want to align all the children of the Stack in the same way, you can use the alignment property of the Stack itself. By default, it's set to be top left, as you have seen in the previous examples. However, you can change it to be centered, as follows:

lib/example_stack_3.dart

```
Stack(
  alignment: Alignment.center,
  children: [
    Container(
      width: 300,
      height: 300,
      color: Colors.red,
    ),
    Container(
      width: 200,
      height: 200,
      color: Colors.green,
    ),
    Container(
      width: 100,
      height: 100,
      color: Colors.blue,
    ),
  ],
),
```

This code will ensure that all the children of the Stack are aligned centrally and will produce a UI that looks like this:

Figure 2.10 – A stack with three colored containers aligned centrally

Next, we'll look at flexible layouts.

Building flexible layouts

While the `Stack` widget arranges widgets on top of each other, you might want to lay down multiple widgets in just one dimension. In the flexible layout model, the children of a `Flex` widget can be laid out in a single direction and can flex their sizes, either growing to fill unused space or shrinking to avoid overflowing the parent.

Row and Column

In Flutter, `Row` and `Column` are `Flex` widgets used for arranging child widgets horizontally and vertically, respectively.

The `Row` widget arranges its children widgets horizontally, in a single row. Each child widget in a row is laid out in the order in which it appears in the children list.

Similarly, the `Column` widget arranges its children widgets vertically, in a single column. Each child widget in a column is laid out in the order it appears in the children list.

Both `Row` and `Column` widgets can have an optional `mainAxisAlignment` property, which defines how the children widgets are positioned along the main axis (horizontal for `Row`, vertical for `Column`). The default value for this property is `MainAxisAlignment.start`, which aligns the children at the start of the main axis. Other possible values for this property include `.center`, `.end`, `.spaceBetween`, and `.spaceAround`.

Both widgets also have an optional `crossAxisAlignment` property, which defines how the child widgets are positioned along the cross axis (vertical for `Row`, horizontal for `Column`). The default value for this property is `CrossAxisAlignment.center`, which centers the children along the cross axis. Other possible values for this property include `.start`, `.end`, and `.stretch`. At first, all of this terminology may sound confusing, but it becomes intuitive quite fast. Here is an example of using the `Row` and `Column` widgets:

lib/example_column_row.dart

```
runApp(
  MaterialApp(
    home: Scaffold(
      body: Column(
        mainAxisAlignment: MainAxisAlignment.center,
        crossAxisAlignment: CrossAxisAlignment.start,
        children: [
          Text('1'),
          Text('2'),
          Row(
            mainAxisAlignment: MainAxisAlignment.spaceAround,
            children: [
```

```
            Text('3'),
            Text('4'),
          ],
        ),
        Text('5'),
      ],
    ),
  ),
 ),
)
```

In the preceding example, we have a `Column` widget with four child widgets. The first two child widgets are `Text` widgets that are arranged vertically in the `Column`. They are laid out vertically beginning from the center of the screen because we have set the `mainAxisAlignment` property of the `Column` to `center`. They are aligned to the left side of the screen (horizontally) because we have set the `crossAxisAlignment` property to `start`. The third child widget is a `Row` widget that arranges two `Text` widgets horizontally with even free spaces between them, as well as around them. This is thanks to the `mainAxisAlignment` property of the `Row` being set to `spaceAround`. The fourth child widget is another `Text` widget arranged vertically in the `Column`. To understand better, let's see the UI that this code produces:

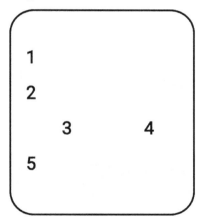

Figure 2.11 – A column with text widgets and a row with various alignments

By default, the `Row` and `Column` widgets will allocate the space needed for each of their children. However, this behavior can be modified using the `Flexible` widget.

Flexible and Expanded

The Flexible widget is a modifier widget that can be used with other widgets to make them flexible within a Row or Column. The Flexible widget can be used to specify how much space a widget should take up within a Row or Column relative to the other children.

The Flexible widget has a flex property that can be used to specify how much of the remaining space should be allocated to the widget. For example, if a row has three children and each child has a flex value of 1, then each child will take up one-third of the available space. If one child has a flex value of 2 and the other two have a flex value of 1, then the first child will take up twice as much space as the other two combined.

For example, consider the following code:

lib/example_flexible_expanded_1.dart

```
runApp(
  MaterialApp(
    home: Scaffold(
      body: Row(
        children: [
          Flexible(
            child: Container(height: 50.0, color: Colors.green),
            flex: 1,
          ),
          Flexible(
            child: Container(height: 50.0, color: Colors.red),
            flex: 2,
          ),
        ],
      ),
    ),
  ),
);
```

In the preceding code, the result of adding all the flex properties is 3. The first child is a Flexible widget with a flex property set to 1, which means it will take up one-third of the remaining space within the Row widget. The second child will then take up two-thirds of the space. This is how it will look:

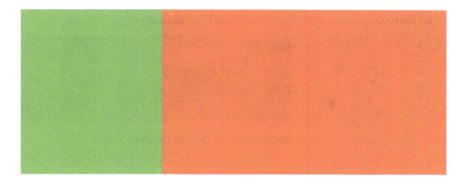

Figure 2.12 – A layout with Flexible and Expanded with various flex properties

Notice that it is possible to combine non-flexible widgets with Flexible widgets. In such a case, the widgets that are not wrapped with a Flexible space will try to allocate their own desired extent and then Flexible will use the space that is left:

lib/example_flexible_expanded_2.dart

```
runApp(
  MaterialApp(
    home: Scaffold(
      body: Row(
        children: [
          Flexible(
            child: Container(height: 50.0, color: Colors.red),
            flex: 1,
          ),
          Container(width: 50.0, height: 50.0, color: Colors.green),
          Flexible(
            child: Container(height: 50.0, color: Colors.blue),
            flex: 1,
          ),
        ],
      ),
    ),
  ),
);
```

As a result of the preceding code, the widgets will be laid out as follows.

Figure 2.13 – A Row with three elements with different widths

In the preceding figure, the second child (green) is a fixed-width container with a width of 50.0 pixels. The first (red) and third (blue) child are Flexible widgets wrapped around a container with a height of 50.0 pixels. The flex property of each Flexible widget is set to 1, and as the total flex count is 2, each Flexible will take up half of the remaining space within the Row widget.

The Flexible widget has a fit parameter, which is set to FlexFit.loose by default. This means that the child of the Flexible can be as large as the available space, but it is also allowed to be smaller than that. However, if we set the fit variable to FlexFit.tight, the child will be forced to take up all the available space. To avoid explicitly setting the fit to tight, we can use a helpful widget called Expanded. The Expanded widget is a specific version of the Flexible widget that sets the fit property to fit FlexFit.tight. Let's compare Flexible and Expanded in an example:

lib/example_flexible_expanded_3.dart

```
runApp(
  MaterialApp(
    home: Scaffold(
      body: Row(
        children: [
          Expanded(
            flex: 1,
            child: Container(
              color: Colors.red,
              height: 100,
              width: 100,
            ),
          ),
          Flexible(
            flex: 1,
            child: Container(
              color: Colors.green,
              height: 100,
              width: 100,
            ),
```

```
            ),
          ],
        ),
      ),
    ),
  );
```

The preceding code will cause the children to be laid out as in the following figure.

Figure 2.14 – A row with one flexible child and another expanded one

In the preceding figure, we have a row with two children. As Flexible and Expanded have flex set to 1, both will try to take up half of the available space. The main difference is that both of their children have the desired width set to 100. The Expanded widget (red) will force the Container to expand to fill the available space, while the Flexible widget (green) will let it take its desired size if it is smaller than the available space.

Solving the overflow problem

The most common issue Flutter developers have is the overflow problem. This mostly happens when using rows and columns when the content inside these widgets exceeds the available space. This can cause the content to be clipped, hidden, or displayed incorrectly, leading to a poor user experience.

Let's consider an example. Suppose you have a row with three children: two Text widgets and a Container widget. If the combined width of these widgets exceeds the width of the device screen, the Text or Container will be clipped or hidden from view. For example, we could write code like this:

lib/example_flexible_expanded_4.dart

```
runApp(
    MaterialApp(
      home: Scaffold(
        body: Row(
          children: [
            Text('First text'),
            Container(
              color: Colors.red,
              width: 1000000,
              height: 100,
```

```
      ),
      Text('Second text')
    ],
  ),
 ),
),
);
```

At first glance, the code looks all right. However, when we render this UI, we will get what is known as an overflow error: the children of the row didn't fit in the available space because the `width` of the `Container` is too big. We can see this on the screen, as the second `Text` is never rendered. Additionally, there is an error in the IDE console.

Figure 2.15 – A layout with an overflowing Container

To solve this problem, we can use the following solutions:

- Wrap the content inside the `Row` or `Column` with an `Expanded` or `Flexible` widget. This will cause the content to take up all the available space inside the `Row` or `Column`. We will omit the wrapping code for running the app, as well as `Scaffold`, and will just focus on the widget we pass in the `body`. For example, see the following code block:

lib/example_flexible_expanded_5.dart

```
Row(
  children: [
    Expanded(child: Text('First text')),
    Expanded(
```

```
        child: Container(
          color: Colors.red,
          width: 1000000,
          height: 100,
        ),
      ),
      Expanded(child: Text('Second text'))
    ],
  )
```

Since all of the children are wrapped in Expanded with the same default flex parameter, the available space is distributed evenly among them, thereby ignoring the desire of the Container to have a width of 1000000. It might not always be the result that you want to achieve, so you might consider other options.

- Use a SingleChildScrollView widget to allow the content inside the row or column to scroll if it exceeds the available space, for example:

lib/example_single_child_scroll_view.dart

```
SingleChildScrollView(
  scrollDirection: Axis.vertical,
  child: Column(
    children: [
      Text('1'),
      Text('2'),
      Text('3'),
      Text('4'),
      Container(
        width: 200,
        height: 1000,
        color: Colors.red,
      ),
      Text('5'),
      Text('6'),
      Text('7'),
      Text('8'),
    ],
  ),
)
```

This code allows all the widgets to be laid out at the size they want, including the `Container`. Since they don't fit on the screen, the whole screen is made scrollable. It looks like this:

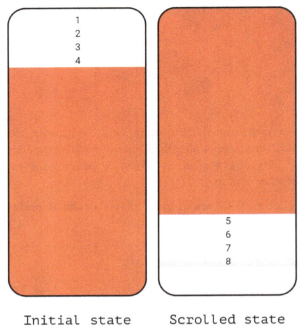

Initial state   Scrolled state

Figure 2.16 – Different states of the SingleChildScrollView

- Use a `Wrap` widget instead of a `Row` or `Column`. The `Wrap` widget automatically wraps its content to the next line when it exceeds the available space. For example, in the following code:

lib/example_wrap.dart

```
Wrap(
  children: [
    Text('First text'),
    Container(
      color: Colors.red,
      width: 100,
      height: 100,
    ),
    Text('Second text'),
    Container(
      color: Colors.green,
```

```
      width: 100,
      height: 100,
    ),
    Text('Third text'),
    Container(
      color: Colors.blue,
      width: 100,
      height: 100,
    ),
  ],
)
```

This code will produce a UI similar to *Figure 2.17*, depending on the screen size. As you can see, when the next widget in the layout didn't have enough space, it was moved to the next line:

Figure 2.17 – A layout of different widgets in a Wrap widget

By implementing one of these solutions, you can avoid the overflow problem with rows and columns in Flutter and provide a better user experience for your app users.

Scrollable items

One of the key features of Flutter is its support for complex scrolling experiences, which enables users to navigate through content that extends beyond the visible screen area. In this section, we will explore the concepts of **ListView**, **GridView**, **CustomScrollView**, and **slivers** in Flutter. We will explore how they can be used to create dynamic and scrollable UIs.

Building scrollable views the easy way

We have just seen the `SingleChildScrollView` widget, which is a scrollable container that can be used to wrap other widgets and enable scrolling when the content exceeds the available space. The `SingleChildScrollView` widget can only have one child widget, which can be any widget that can fit within the constraints of the available space. When the content exceeds the available space, the `SingleChildScrollView` widget will automatically enable scrolling, allowing the user to navigate through the content.

The `SingleChildScrollView` widget is an easy and convenient way to add scrolling capabilities to a single child widget. However, it may not be suitable for more complex layouts. For example, if you have multiple child widgets that need to be scrolled together, you may need to use a different type of scrolling widget.

One issue that you can run into by using `SingleChildScrollView` is that when you wrap any children inside the `Column` or `Row` in an `Expanded` widget, you will encounter an unbounded height or width exception. This will throw something along the lines of "RenderFlex children have non-zero flex but incoming height constraints are unbounded." This happens because the `SingleChildScrollView` parent already assumes infinite height due to scrolling capabilities, yet the `Expanded` child wants to be as big as possible, which in our case is infinite height. This leads to a sizing conflict, which can't be resolved by using these widgets. For example, this code would produce the mentioned error:

lib/example_unbound_height.dart

```
runApp(
  MaterialApp(
    home: Scaffold(
      body: SingleChildScrollView(
        scrollDirection: Axis.vertical,
        child: Column(
          children: [
            Text('1'),
            Text('2'),
            Text('3'),
            Text('4'),
            Expanded(
```

```
        child: Container(
          width: 200,
          height: 1000,
          color: Colors.red,
        ),
      ),
      Text('5'),
      Text('6'),
      Text('7'),
      Text('8'),
    ],
  ),
),
),
),
);
```

If you have a list of homogenous items, there is an easy way to display them as a list or as a grid. We have seen how to do that with a ListView widget in *Chapter 1* and with a GridView widget earlier in this chapter, when we talked about building responsive layouts according to information from MediaQuery. While these widgets cater to a lot of use cases, we still can sometimes run into even more complicated designs when there are many moving yet different parts. In Flutter, we have a whole system of widgets dedicated specifically to that. They are called **slivers**.

The secret of slivers

Slivers are a special type of widget that are used in conjunction with scrollable containers to enable more complex scrolling behaviors. Slivers essentially comprise a type of flexible space that can adapt itself as the user scrolls. You can add slivers into a CustomScrollView to create scrollable areas that contain various types of widgets, such as lists, grids, and cards. You can also define custom scrolling behaviors and animations.

Some of the most commonly used sliver widgets include the following:

- SliverAppBar: This is a sliver widget that provides an app bar that can expand and contract as the user scrolls. The sliver app bar can also float at the top of the screen and display a flexible space.

- SliverList: This is a sliver widget that displays a list of items. The SliverList is designed to work efficiently with large lists and can be used to build scrolling lists that are optimized for performance.

- SliverGrid: This is a sliver widget that displays a grid of items. The SliverGrid can be used to create scrolling grids that can be customized with various properties, such as the number of columns and the size of each grid item.

- `SliverToBoxAdapter`: This is a sliver widget that allows you to insert a non-scrollable widget into a `CustomScrollView`. It is important to note that all children of the `CustomScrollView` must be slivers. If you try to insert a regular widget instead of a sliver, such as a simple `Container`, you will get an error that says **A RenderViewport expected a child of type RenderSliver but received a child of type RenderConstrainedBox**. To avoid that, wrap any widget that is not a sliver and can't be represented with a sliver analog in a `SliverToBoxAdapter`.

Here's an example of how to create a `CustomScrollView` with a `SliverAppBar` and a `SliverList`:

lib/example_sliver_1.dart

```
runApp(
  MaterialApp(
    home: Scaffold(
      body: CustomScrollView(
        slivers: [
          SliverAppBar(
            title: Text('My App'),
          ),
          SliverList(
            delegate: SliverChildBuilderDelegate(
              (BuildContext context, int index) {
                return ListTile(
                  title: Text('Item $index'),
                );
              },
              childCount: 100,
            ),
          ),
        ],
      ),
    ),
  ),
);
```

In the preceding example, we created a `CustomScrollView` that contains a `SliverAppBar` and a `SliverList`. The `SliverAppBar` provides an app bar that can expand and contract as the user scrolls and the `SliverList` displays a list of items.

When rendered, it will look like this:

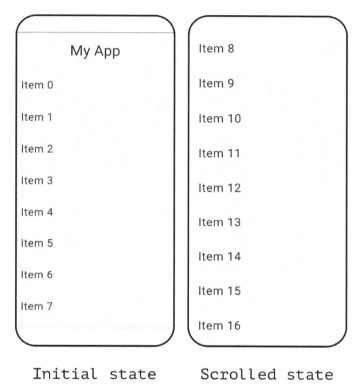

Figure 2.18 – A basic layout with slivers

In addition to the built-in sliver widgets, Flutter also provides a way to create custom sliver widgets. Custom slivers can be created by extending the `SliverPersistentHeaderDelegate` class and overriding its methods to customize the behavior of the sliver. This provides a high degree of flexibility when creating complex scrolling behaviors. Let's see an example of how to use `SliverPersistentHeaderDelegate` in Flutter:

lib/example_sliver_2.dart

```
class MySliverHeaderDelegate extends SliverPersistentHeaderDelegate {
  final double maxHeight;
  final double minHeight;
  final Widget child;

  MySliverHeaderDelegate({
    required this.maxHeight,
```

```
    required this.minHeight,
    required this.child,
  });

  @override
  Widget build(BuildContext context, double shrinkOffset,
                bool overlapsContent) {
    return SizedBox.expand(child: child);
  }

  @override
  double get maxExtent => maxHeight;

  @override
  double get minExtent => minHeight;

  @override
  bool shouldRebuild(SliverPersistentHeaderDelegate oldDelegate) {
    return true;
  }
}
```

In the preceding example, we have defined a custom `MySliverHeaderDelegate` that extends `SliverPersistentHeaderDelegate`. This delegate takes in a `maxHeight` and `minHeight` of type `double`, and a `child` of type `Widget`.

The `build` method returns a `SizedBox` that expands to fill the available space and contains the child widget. The `maxExtent` and `minExtent` methods return the corresponding heights and the `shouldRebuild` method always returns `true`, indicating that the header should be rebuilt if the delegate changes. For example, let's say that we want a blue `Container` with the text **My Header** to expand from height `100` to `200` when scrolled down, and to collapse from `200` to `100` when scrolled up. To use this custom header delegate, we can create a `CustomScrollView` with a `SliverPersistentHeader`:

lib/example_sliver_2.dart

```dart
class MyScrollView extends StatelessWidget {
  @override
  Widget build(BuildContext context) {
    return Scaffold(
      body: CustomScrollView(
        slivers: [
          SliverPersistentHeader(
            delegate: MySliverHeaderDelegate(
              maxHeight: 200,
              minHeight: 100,
              child: Container(
                color: Colors.blue,
                child: Center(
                  child: Text('My Header'),
                ),
              ),
            ),
            pinned: true,
          ),
          SliverList(
            delegate: SliverChildBuilderDelegate(
              (BuildContext context, int index) {
                return ListTile(
                  title: Text('Item $index'),
                );
              },
              childCount: 100,
            ),
          ),
        ],
      ),
    );
  }
}
```

This code will produce the following UI:

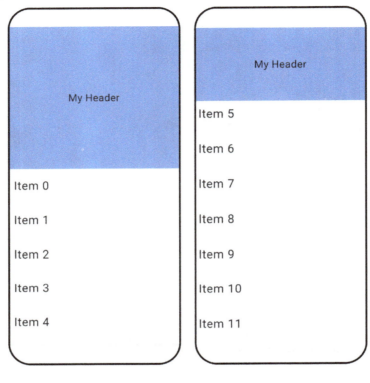

Initial state   Scrolled state

Figure 2.19 – A layout with a custom sliver

In the preceding example, we created a `CustomScrollView` that contains a `SliverPersistentHeader` with our custom delegate. The `SliverPersistentHeader` is set to be pinned, which means that it will remain visible at the top of the screen even as the user scrolls.

The `CustomScrollView` also contains a `SliverList` that displays a list of items. By combining these two slivers, we can create a scrolling interface that contains a custom header that remains visible as the user scrolls through the list.

Other layouts

While flexible and scrollable widgets can feel like enough to build our desired UI, there might be some scenarios in which we might need more complex layouts. For those cases, we can make use of the `Wrap` and `Flow` widgets or build our custom own solution. The `Wrap` widget is a layout widget that can be used to wrap a set of child widgets with specific alignment and spacing. The `Wrap` widget is similar to the `Row` and `Column` widgets, but it provides more flexibility by allowing child widgets to flow onto multiple lines as needed. We have already seen it in action earlier in this chapter when we discussed how to handle overflow errors.

The flow layout

The `Flow` widget is a layout widget that can be used to create custom layouts by manually positioning child widgets within the available space. The `Flow` widget allows you to specify the position of each child widget using a set of transformation matrices.

Let's say that we want to create a custom layout wherein child widgets are positioned along the circumference of a circle. We could use the `Flow` widget to achieve this. Here is an example of how the `Flow` widget can be used to create a circular layout:

lib/example_flow.dart

```
runApp(
  MaterialApp(
    home: Scaffold(
      body: Flow(
        delegate: CircleLayoutDelegate(),
        children: [
          Container(w: 50.0, h: 50.0, color: Colors.red),
          Container(w: 50.0, h: 50.0, color: Colors.green),
          Container(w: 50.0, h: 50.0, color: Colors.blue),
          Container(w: 50.0, h: 50.0, color: Colors.yellow),
          Container(w: 50.0, h: 50.0, color: Colors.purple),
          Container(w: 50.0, h: 50.0, color: Colors.pink),
          Container(w: 50.0, h: 50.0, color: Colors.orange),
        ],
      ),
    ),
  ),
);
```

Instead of laying out children in a boring horizontal or vertical line, we now lay them out in a circle like this:

Figure 2.20 – A Flow layout with children laid out in a circle

In the preceding example, we have defined a custom layout strategy by creating a `CircleLayoutDelegate` class that extends the `FlowDelegate` class. The `CircleLayoutDelegate` class overrides the `paintChildren()` method to specify the position of each child widget using a set of transformation matrices. The `Flow` widget will use the `CircleLayoutDelegate` class to position the child widgets within the available space. We won't be looking into the code of `CircleLayoutDelegate` in this chapter, as it mostly just deals with mathematical calculations. If you're curious, you can find it in the sample code at `lib/example_flow.dart`.

Building your own layout

The CustomMultiChildLayout widget in Flutter is a powerful and flexible way to create custom layouts by manually positioning multiple child widgets within the available space. It is a layout widget that takes a delegate as a parameter, which is responsible for positioning the child widgets. This delegate must extend the MultiChildLayoutDelegate class, which defines methods for measuring and positioning child widgets. The delegate is responsible for laying out the child widgets and returning their positions as Size and Offset objects.

The MultiChildLayoutDelegate class has several methods that can be overridden to define the layout strategy. These include the following:

- getSize(BoxConstraints constraints): This method is responsible for calculating the size of the layout based on the constraints provided by the parent widget.

- performLayout(Size size): This method is responsible for laying out the child widgets within the available space.

- hasChild(int index): This method is responsible for determining whether a child widget exists at the specified index.

- getFirstChildKey() and getLastChildKey(): These methods are responsible for returning the keys of the first and last child widgets, respectively.

The CustomMultiChildLayout widget is useful in situations where the standard layout widgets, such as Row and Column, are not flexible enough to achieve the desired layout.

Let's say that we want to create a custom layout wherein child widgets are arranged in a grid layout. We could use the CustomMultiChildLayout widget to achieve this.

For example, let's say that we want to lay out our children in a 3x3 grid. If there are less than 3 items in a row, the last item should occupy all the available space.

Figure 2.21 – A layout with a custom MultiChildLayoutDelegate

Here is an example of how the `CustomMultiChildLayout` widget can be used to create such a grid layout. First, we need to override the `MultiChildLayoutDelegate`:

lib/example_custom_multichild_layout.dart

```
class CustomGridLayoutDelegate extends MultiChildLayoutDelegate {
 @override
 void performLayout(Size size) {
   int count = 4;

   final double itemWidth = size.width / 3;
   final double itemHeight = size.height / 3;

   for (int i = 0; i < count; i++) {
     layoutChild(i, BoxConstraints.loose(size));

     final double xPos = (i % 3) * itemWidth;
```

```
      final double yPos = (i ~/ 3) * itemHeight;

      positionChild(i, Offset(xPos, yPos));
    }
  }

  @override
  bool shouldRelayout(covariant MultiChildLayoutDelegate oldDelegate)
  => true;
}
```

In the preceding example, we have defined a custom layout strategy by creating a `CustomGridLayoutDelegate` class that extends the `MultiChildLayoutDelegate` class. The `CustomGridLayoutDelegate` class overrides the `performLayout()` method to position each child widget within the available space, having at most 3 children in a row.

Now, in order to display a grid of such items, we could do the following:

lib/example_custom_multichild_layout.dart

```
void main() {
  runApp(
    MaterialApp(
      home: Scaffold(
        body: CustomMultiChildLayout(
          delegate: CustomGridLayoutDelegate(),
          children: [
            GridItem(id: 0, color: Colors.red),
            GridItem(id: 1, color: Colors.green),
            GridItem(id: 2, color: Colors.blue),
            GridItem(id: 3, color: Colors.yellow),
          ],
        ),
      ),
    ),
  );
}

class GridItem extends StatelessWidget {
  final Color color;
  final int id;

  const GridItem({super.key, required this.color, required this.id});
```

```
  @override
  Widget build(BuildContext context) {
    return LayoutId(
      id: id,
      child: Container(
        color: color,
      ),
    );
  }
}
```

We have defined four child widgets called `GridItem`, each wrapped in a `LayoutId` parent. This is required to pass an `id`, which is then retrieved in the `for` loop via the `i` variable. We have currently hardcoded the number of items to be 4, but this can easily be refactored to have a variable number of children. Those children are passed to the `CustomMultiChildLayout` widget, which uses our `CustomGridLayoutDelegate` as the delegate.

When the `CustomMultiChildLayout` widget is built, it passes its size to the delegate's `performLayout()` method. The `CustomGridLayoutDelegate` then calculates the size of each child widget and positions them in a 3x3 grid layout, with a spacing of 0 between each widget. If there are fewer than 3 children in a row, the last child occupies all of the space that is left.

Ensuring accessibility in Flutter apps

Accessibility is an important aspect of building apps that are inclusive for all users, regardless of their ability or disability. Flutter provides several widgets and tools to make it easy to build accessible apps.

> **What is accessibility?**
>
> Accessibility is the practice of designing and building apps that can be used by people with disabilities. It involves making sure that users with disabilities can interact with your app using assistive technologies such as screen readers, switch devices, and keyboard-only navigation.
>
> Some common disabilities that affect how people interact with digital products include visual, auditory, cognitive, and motor impairments. For example, someone who is visually impaired may rely on a screen reader to navigate an app, while someone who is physically impaired may use a switch device to interact with the app.

Getting to know accessibility widgets in Flutter

Flutter provides several widgets that are designed to be accessible out of the box. These widgets include the following:

- `Semantics`: This widget is used to provide accessibility information about other widgets in your app. For example, you can use `Semantics` to specify the role of a widget (such as `button` or `text field`), describe its purpose, or provide a hint to assistive technologies about how to interact with it.

- `ExcludeSemantics`: This widget is used to exclude a widget from the accessibility tree. It can be useful in situations where a widget should not be announced by assistive technologies, such as a decorative image.

- `FocusNode`: This widget is used to manage focus within your app. Focus is important for users who rely on keyboard-only navigation, as it allows them to navigate through your app using the *Tab* key.

- `FocusScope`: This widget is used to create a scope for focus within your app. It can be useful for grouping related widgets together and ensuring that the focus stays within a particular area of the app.

- `MergeSemantics`: This widget is used to merge the accessibility information of multiple widgets into a single `Semantics` node. It can be useful for grouping related widgets together and providing a single, coherent accessibility experience.

Next up, let's look at how to style our words on screen.

Font size and color contrast

Flutter app users can adapt their font sizes on both Android and iOS through the system settings. Flutter text widgets automatically adjust font sizes according to these settings.

As a developer, it's important to ensure that your app's layout accommodates larger font sizes by leaving enough space to render all content. To do this, you can test your app on a small-screen device with the largest font setting enabled to ensure that all parts of the app are still visible and usable.

Color contrast ratio is also an important aspect of accessibility in a Flutter app because it affects the legibility and usability of the app for users with visual impairments or color vision deficiencies.

In general, a contrast ratio of at least 4.5:1 is recommended between the foreground (text or graphics) and background colors of an interface. This helps ensure that users with low vision or color blindness can distinguish between different elements on the screen. Additionally, it helps prevent eye strain for all users.

Lastly, **dark mode**, which is often viewed as an appealing extra for enthusiasts, goes beyond aesthetics and serves as a significant accessibility feature that some might overlook. This mode is especially beneficial for individuals with certain visual impairments, as bright colors and extensive white spaces can be discomforting to them. Black text on a white background can also pose challenges to people with conditions such as dyslexia or Irlen Syndrome. Similar to other accessibility tools, dark mode offers each user the flexibility to interact with applications in the way that suits them best. In Flutter, this can easily be achieved with theming. For example, you might specify ThemeData for the theme and darkTheme parameters, as well as any other theme parameters of the MaterialApp widget.

Dev tooling for accessibility

The Flutter framework also provides the tools to verify whether your app meets the requirements to be considered accessible. While we will cover testing in *Chapters 11* and *12*, we can have a look at how we can use AccessibilityGuideline to test our app's accessibility.

The framework provides four main guidelines:

- **androidTapTargetGuideline**: Verifies whether tappable nodes have a minimum size of 48 by 48 pixels
- **iOSTapTargetGuideline**: Verifies whether tappable nodes have a minimum size of 44 by 44 pixels
- **textContrastGuideline**: Provides guidance for text contrast requirements, as specified by WCAG
- **labeledTapTargetGuideline**: Enforces that all nodes with a tap or long press action also have a label

To check whether our app meets any of these guidelines, we can create a test that verifies this. In Flutter, you could write a widget test like this:

```
testWidgets('HomePage meets androidTapTargetGuideline',
    (WidgetTester tester) async {
  final SemanticsHandle handle = tester.ensureSemantics();
  await tester.pumpWidget(const MaterialApp(home: HomePage()));
  await expectLater(tester,
  meetsGuideline(androidTapTargetGuideline));
  handle.dispose();
});
```

In the preceding test widget, we verified that the HomePage meets the androidTapTargetGuideline.

What is a widget test?

Don't worry if you don't know yet how to implement testing in your application. We will cover that in *Chapter 11*. Make sure to review this section once you feel more comfortable with widget tests.

Manual review

As you prepare your app for release, there are several considerations to keep in mind. Some of these include the following:

- **Active interactions**: Ensure that all active interactions have a purpose. If a button is pushed, it should perform an action. If there is currently a no-op callback for an `onPressed` event, consider updating it to show a `SnackBar` that explains which control was just pushed.

- **Screen reader testing**: Test your app with TalkBack (Android) and VoiceOver (iOS) to ensure that the screen reader can accurately describe all controls on the page and that the descriptions are intelligible.

- **Contrast ratios**: We encourage you to have a contrast ratio of at least 4.5:1 between controls or text and the background. Images should also be reviewed for sufficient contrast, with the exception of disabled components.

- **Context switching**: Nothing should automatically change the user's focus while they are entering information. Widgets should generally avoid changing the user's context without some sort of confirmation action.

- **Tappable targets**: All tappable targets should measure at least 48x48 pixels to ensure ease of use.

- **Errors**: Important actions should be able to be undone. Fields that show errors should suggest a correction if possible.

- **Color vision deficiency testing**: Controls should be usable and legible in colorblind and grayscale modes.

- **Scale factors**: The UI should remain legible and usable at very large-scale factors for text size and display scaling.

Accessibility is an important aspect of building apps that are inclusive for all users. Flutter provides several widgets and tools to make it easy to build accessible apps and it is our job to give priority to accessibility in our app development process.

Summary

In this chapter, we learned about the layout system in Flutter. We learned how to create complex and flexible layouts that are accessible to all users. The layout algorithm in Flutter is based on the *"Constraints go down. Sizes go up. Parent sets the position"* rule. This means that the parent widget passes constraints to its children, which then determine their size. The parent then positions the children based on their size. However, widgets are not aware of their own position. The final size and position of a widget depends on its parent. Despite these limitations, Flutter's layout algorithm is very efficient and can be used to create responsive UIs that look great on any screen size.

We learned how to obtain information about the device's screen size and orientation to adjust the UI based on these parameters. We also covered the use of widgets such as Stack or Row to control the position and size of other widgets. We learned that accessibility is an important aspect of building apps that are inclusive for all users. As we have learned in this chapter, Flutter provides several widgets and tools to make it easy to build accessible apps. The importance of creating responsive UIs will only continue to grow as Flutter continues to evolve and support more platforms, making it essential for developers to master these strategies. These lessons are important for creating UIs that are adaptable to different devices and screen sizes, fast, responsive, and accessible to all users.

In *Chapter 3*, we will dig into how to manage the state of our app, as well as consider which challenges it presents. We will also start applying everything that we have learned so far with a new example project that we will start building in the following chapters.

Get this book's PDF version and more

Scan the QR code (or go to packtpub.com/unlock). Search for this book by name, confirm the edition, and then follow the steps on the page.

Note: Keep your invoice handy. Purchases made directly from Packt don't require an invoice.

Part 2: Connecting UI with Business Logic

In this part, you will learn how to take your apps beyond a beautiful UI by efficiently connecting them to business logic. We will explore which patterns are used in the Flutter framework and how they impact our choice of state management solutions. We will look at various approaches, from vanilla state management to third-party tools, and explore their benefits and limitations. This experience will help us make informed decisions. Moreover, we will learn how to implement efficient navigation in our applications.

This part includes the following chapters:

- *Chapter 3, Vanilla State Management*
- *Chapter 4, State Management Patterns and Their Implementations*
- *Chapter 5, Creating Consistent Navigation*

3
Vanilla State Management

Beautiful, responsive applications capture the user's attention and leave a positive impression about your product. However, there is much more going on behind what the user sees as they navigate and interact with your app. Button taps, network requests, saving data, showing progress, conveying errors, and delivering notifications are just a tiny fraction of what's happening in a typical app. Handling all of this flawlessly for the user greatly depends on the **state management** techniques that are used under the hood.

In this chapter, we will learn what state management is and what challenges it presents, what tools the Flutter framework offers developers to tackle these tasks, and how to use them efficiently without introducing unwanted bugs.

We will be working on a sample eCommerce application called Candy Store to gather knowledge.

By the end of this chapter, you will feel confident in your ability to tackle state management challenges by using only the tools available in the Flutter SDK out of the box, observe their peculiarities and limitations, and gain the fundamental knowledge required to understand more advanced techniques.

In this chapter, we're going to cover the following main topics:

- What is state and why do you need to manage it?
- Getting to know our Candy Store app
- Managing state the vanilla Flutter way
- Passing around dependencies via InheritedWidget
- Interacting with BuildContext in the right place at the right time

Technical requirements

If this is not your first Flutter app, then you probably already have everything you need installed.

Otherwise, you will need to install the following:

- An IDE of your choice that supports Flutter, such as Android Studio or VS Code
- The Flutter SDK

You can find all of the code required for this chapter here: `https://github.com/PacktPublishing/Flutter-Design-Patterns-and-Best-Practices/tree/master/CH03`.

You can find full versions of the source code snippets that we will be examining here:

- `https://github.com/flutter/flutter/blob/master/packages/flutter/lib/src/widgets/framework.dart`
- `https://github.com/flutter/flutter/blob/master/packages/flutter/lib/src/foundation/change_notifier.dart`
- `https://github.com/flutter/flutter/blob/master/packages/flutter/lib/src/widgets/media_query.dart`

What is state and why do you need to manage it?

You can probably guess that **state** is something integral to app development. You can further understand the importance of state in Flutter by the fact that the two main types of widgets contain the word state in their title: `StatelessWidget` and `StatefulWidget`. To have a state or not to have, that is the question. But first, let's figure out what state is.

What is state?

State is the representation of data at any given time. It can also refer to the status of user interface elements. The state can change due to many reasons, such as user interaction, updates from data sources, or reactions to animation ticks. State can be anything between something as small as the index of the selected tab or a string of text on the screen, and a full-blown user profile showing all its details. However, this broad definition of state can sometimes cause more confusion than clarity. The topic of state, and specifically its management, is notorious in the Flutter community. This isn't much of a surprise since the state, in some sense, is the essence of the app. It conveys the current state of things to the user through a visual user interface and translates data into something that the user can consume and interact with.

Understanding the difference between ephemeral and app state

The official Flutter documentation (https://docs.flutter.dev/data-and-backend/state-mgmt/ephemeral-vs-app) has a dedicated section on state management. The official documentation is intentionally quite vague in describing what state is and how to manage it because, as with many things in programming, it depends on many factors: your app, your knowledge, and your preferences. That being said, there is a useful notion that can help us tackle state – being able to separate state into **ephemeral state** and **app state**.

The Cambridge Dictionary defines the word *ephemeral* as *"lasting for only a short time, synonym to short-lived"* (https://dictionary.cambridge.org/dictionary/english/ephemeral). So, the first difference between ephemeral and app state is the *lifetime* or *persistence* of the state. To understand how to handle the state according to its lifetime, we need to answer the following questions:

- How long is this state relevant?
- Where does this state need to be stored?
- Does this state have value only while the user is on the current page or does the user need to save this information even if the app is suddenly closed?

These questions also highlight the second difference – the *scope* of the state:

- In how many places should this state be accessible?
- Can it be self-contained inside of a single widget?
- Should this data be available from various pages of the application?

It is also important to understand that answers to these questions will vary from case to case as it highly depends on the functionality and business logic of the specific application. For some, it is OK if the text from TextField is cleared once the app is closed; for others, it is crucial to save it so that the user can continue where they left off. As we discussed earlier – it depends.

The difference between ephemeral state and app state is summarized in the following table:

	Ephemeral State	**App State**
Contains important business logic	Usually no	Often yes
Can survive app restart	No	Sometimes yes
Accessible to more than one widget	No	Often yes

	Ephemeral State	App State
Is relevant across several pages	No	Can be
Is backed up by more persistent storage (database, server, and so on)	No	Can be
Lifetime	Short-lived, usually while the widget is visible on the screen	Anything from short-lived to persistent information in long-term storage
Scope	Self-contained within the host widget	Can vary from self-contained to the whole app level

Table 3.1 – Summary of the difference between ephemeral and app state

Although I agree with the idea, I'm not exactly a fan of the wording *ephemeral* versus *app* state: *ephemeral* is still a type of app state; it just has a shorter lifespan and a tighter scope. So, when it comes to *state management* in this book, we will assume the management of the lifetime and scope of the application's state. In a way, this is also managing the application in general since everything is, in some sense, app state.

Since the definition of *state* is quite broad, it means that the problems related to it will be quite wide-ranging as well. To build our toolbox from the ground up and explore the pros, cons, and issues that are present in various approaches, we will learn from practice. We will build a sample app and refactor it step by step as we learn new things along the way.

Getting to know our Candy Store app

The app that we will be building throughout several of the following chapters will be a typical eCommerce app. You can check out the initial code for this chapter by going to the folder CH03/initial, and you can view the final result, as well as step-by-step refactoring, via the git commit history in CH03/final. Now, let's take a look at the wireframes of the app that we will be working with, starting with MainPage and ProductsPage:

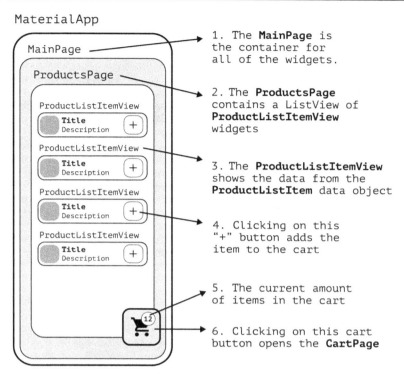

Figure 3.1 – The Candy Store app wireframe for MainPage and ProductsPage

The initial version of the app will consist of only two functional pages and its root will be `MaterialApp`. The app will be made up of the following components:

- `MainPage`: This is the wrapper page for the app. `ProductsPage` is its child, and the **Cart** button is also part of it.

- `ProductsPage`: This is a simple `ListView` that builds `ProductListItemView`. Each `ProductListItemView` is backed up by a data object called `ProductListItem`, which contains the title, description, image, and other product information. On clicking the + button of `ProductListItemView`, this item gets added to the cart.

- The **Cart** button: The total amount of items in the cart is shown via the **Cart** button. There can be several instances of one product, but this number shows all of them. Upon clicking the **Cart** button, `CartPage` will open.

With that, we have reviewed the various components of ProductsPage. Now, let's take a look at CartPage. The wireframe of CartPage looks like this:

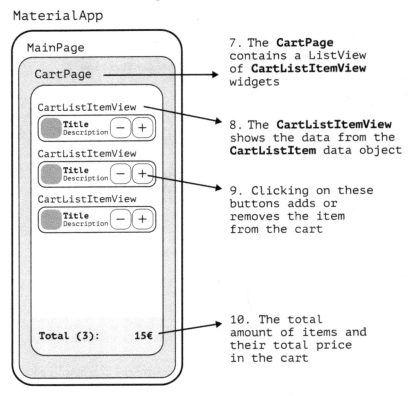

Figure 3.2 – The Candy Store app wireframes for CartPage

The CartPage class consists of the following elements:

- CartPage: This is built in a similar way to ProductPage, except that instead of ProductListItemView, it shows CartListItemView. The total amount of items and their total price is shown at the bottom of CartPage. It is updated dynamically as the items get added and removed from the cart.

- CartListItemView: This is backed up by a CartListItem data object. It contains information about the product, as well as how many instances of this product are present in the cart. We can remove and add the same type of items to the existing cart. It also updates the list on MainPage so that the count on the **Cart** button is updated when we return to it.

Let's also review the Candy Store app from the widget tree's perspective:

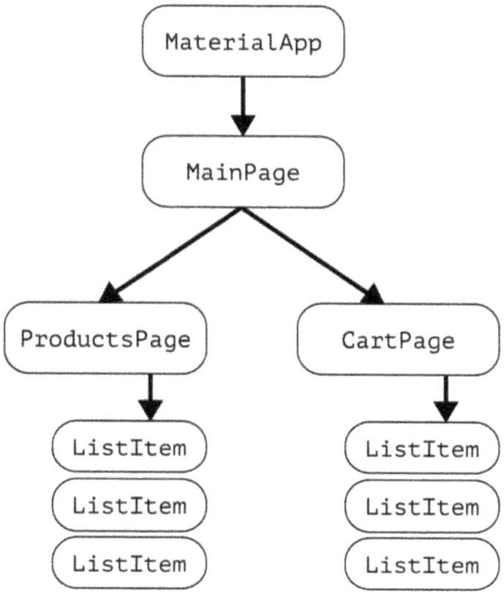

Figure 3.3 – The Candy Store app widget tree

This is a simplified version of the widget tree that includes only the widgets that are relevant to demonstrate the data flow. Note that `ProductsPage` and `CartPage` are independent widget stacks; this knowledge will be important later on. Now that we know what we're going to build, let's get started.

Managing state the vanilla Flutter way

While the debate over the *best* state management solution is never-ending, it is important to note that all of the third-party libraries are built on top of the existing Flutter framework. In software development, the term *vanilla* is used to describe something that is used without modification. In our case, the vanilla Flutter framework refers to the raw framework without any dependencies. Understanding the fundamentals of how popular libraries are built can help inform how we work with libraries, solutions, and our existing frameworks. Sometimes, the good old `setState` is enough.

Lifting the state up

To implement our Candy Store app, the first thing we must consider is how to access the list of items in the cart. Remember that we can add items to the cart from the products page and manage the cart from the cart page itself. To do this, we need a way to share cart data between these two independent pages.

Although we could technically achieve this using a global or static variable, these approaches are generally considered bad practice in programming for several reasons:

- **Global scope**: These variables are available from anywhere in your code. As the code base grows, it becomes increasingly harder to track the places where those variables are modified, which can lead to nasty and hard-to-trace bugs.

- **Tight coupling**: Abstraction and encapsulation are two of the main pillars of object-oriented programming. Following those principles makes the code more maintainable since it's less coupled, and potential changes can be done in one place instead of all over the place. This makes swapping implementations less painful.

- **Testability**: Having a shared state makes tests dependent on each other, which quickly leads to unstable tests that add more problems than they solve.

We will delve into these issues in much greater detail in *Part 3* and *Part 4* of this book. With these anti-patterns out of the way, let's address the problem at hand.

Let's consider the following aspects:

- Both pages need the cart items list.
- We are working with a tree data structure (of widgets).

In this case, it probably makes sense to have a common parent for those children and let the parent manage the cart items list. The parent can store the list in its own state and, when required, pass that list to its children.

This parent will be MainPage, the code for which is as follows:

lib/main_page.dart

```
class MainPage extends StatefulWidget {
  const MainPage({super.key});

  @override
  State<MainPage> createState() => _MainPageState();
}

class _MainPageState extends State<MainPage> {
  // The Map key is the id of the CartListItem.
  // We will use a Map data structure because
  // it is easier to manage the addition & removal of items.
  final Map<String, CartListItem> items = {};

  @override
```

```
  Widget build(BuildContext context) {...}
}
```

In the preceding code, we have added a `cartItemsMap` field, which will be responsible for storing the items in the cart. So far, so good. However, an empty cart is useless, so let's create a way to add and remove items from it. To do that, we will create two methods, `addToCart` and `removeFromCart`, in `MainPage`:

lib/main_page.dart

```
class _MainPageState extends State<MainPage> {
  final Map<String, CartListItem> items = {};

  // ...

  void addToCart(ProductListItem item) {
    // Add `item` to items
  }

  void removeFromCart(CartListItem item) {
    // Remove `item` from the items
  }
}
```

The actual implementation details of these methods in the sample app are irrelevant at this point. We only need to understand that we manipulate the items directly inside of them and keep the reference to the map inside `_MainPageState`.

`MainPage` itself doesn't call these methods. We can add and remove items from `CartPage` and `ProductsPage`. We can pass those methods as callbacks to their constructors using the following code:

lib/products_page.dart

```
class ProductsPage extends StatefulWidget {
  final Function(ProductListItem) onAddToCart;
  // ...
}
```

lib/cart_page.dart

```
class CartPage extends StatefulWidget {
  final List<CartListItem> items;
  final Function(CartListItem) onRemoveFromCart;
  final Function(CartListItem) onAddToCart;
```

```
// ...
}
```

In the preceding code, we pass the onAddToCart callback to ProductsPage as an argument. We also pass the onAddToCart callback to CartPage, along with onRemoveFromCart and the list of CartListItem values to be displayed.

Because we need to share our list of cart items among several pages, it means that none of those pages can be responsible for storing and manipulating this list. So, what we must do is lift the state to the parent of those pages – to MainPage – and pass down the callbacks that manipulate the cart items list to the children pages – ProductsPage and CartPage. This is called **lifting state up** and it is a very common pattern for dealing with shared state.

If we test our app now, we will notice that the products page works perfectly fine as the number of cart items increases. However, if we do the same from the cart page, nothing changes. This is because we have opened CartPage via Navigator and created a new, independent widget tree on the navigation stack. Therefore, if we want to see updates on CartPage, we need to maintain its internal state and propagate the changes back to the parent.

The code to update the cart could look something like this:

lib/cart_page.dart

```
class _CartPageState extends State<CartPage> {
  List<CartListItem> _items = [];

  @override
  void initState() {
    super.initState();
    _items = widget.items;
  }

  // ...

  void _removeFromCart(CartListItem item) {
    setState(() {
      // Remove `item` from _items, same logic as in MainPage
    });
    widget.onRemoveFromCart(item);
  }

  void _addToCart(CartListItem item) {
    setState(() {
      // Add `item` to the _items, same logic as in MainPage
```

```
      });
      widget.onAddToCart(item);
   }
 }
```

In the preceding code, we keep an internal list of `CartListItem` called `_items`. In `initState`, we assign the list that was passed from the widget to the internal `_items`. Then, every time we need to remove or add an item, we manipulate the internal `_items` field, and then call the callback of the parent.

To better understand how the data flows in this case, let's take a look at the following diagram:

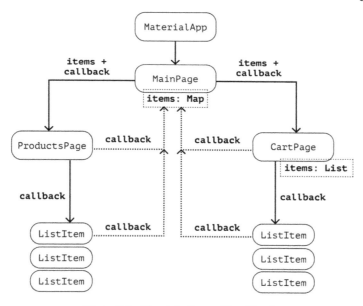

Figure 3.4 – The data flow with callbacks

Some important things to note here are that both `MainPage` and `CartPage` maintain their own collections of `items`, and that we pass a lot of callbacks back and forth.

While this may work, this approach is flawed. Even in a simple scenario, it requires a lot of error-prone boilerplate code. For instance, suppose we want to enable users to view product details from the cart page. The details appear on top, and the user can add the item to the cart again. However, there can be additional constraints, such as the availability of a limited number of items or a specific limit on the number of items per buyer. As the logic becomes more complex, managing it with a simple stateful widget can quickly get out of hand.

Thankfully, Flutter has us covered and offers several tools out of the box. Let's explore them.

Understanding the Observer pattern

Discussing design patterns often involves mentioning the Gang of Four and their influential book, *Design Patterns: Elements of Reusable Object-Oriented Software*. One of the patterns described in this book that is ubiquitous in software development is a behavioral pattern called the **Observer pattern**.

The idea behind this pattern is that there is something, known as the **observable**, that someone called the **observer** wants to observe. For instance, let's say we want to react to changes in network status by displaying a persistent SnackBar whenever there is no internet connection. We will remove that SnackBar when the connection is restored.

Here's a summary of the pattern:

- An observable has some state that can change.
- An observer can receive notifications when the state of the observable changes.
- An observable can have multiple observers.
- The observable's task is to notify its observers, and the observer's task is to react accordingly.

This is a fundamental concept of reactive programming that we will use extensively in this book since declarative and reactive programming go hand in hand. We discussed declarative programming in the *Understanding the difference between declarative and imperative UI design* section in *Chapter 1*; we will discuss reactive paradigm in more detail in *Chapter 4*.

Implementing an observable using the Listenable class

Just like we have the base `Widget` class from which all kinds of widgets inherit, we have a base `Listenable` class, from which all kinds of listenables inherit. You can find the source code for this part of this chapter here: `https://github.com/flutter/flutter/blob/master/packages/flutter/lib/src/foundation/change_notifier.dart`. The interface of the `Listenable` class is very simple:

```
abstract class Listenable {
  const Listenable();

  factory Listenable.merge(List<Listenable?> listenables)
     = _MergingListenable;

  void addListener(VoidCallback listener);

  void removeListener(VoidCallback listener);
}
```

The preceding code is presented without any modifications for demonstration purposes. Here's what it contains:

- A `merge` factory constructor that simply merges several listenables. This is not relevant to our current discussion.

- Two methods for removing and adding listeners.

This is the core concept. The core classes are intentionally kept vague, with only the minimum properties required, so that implementers or overriders can handle more specific cases. This is also why there is a `ValueListenable` class in the core framework. Let's take a look at the implementation of `ValueListenable`:

```
abstract class ValueListenable<T> extends Listenable {
  const ValueListenable();

  T get value;
}
```

The preceding class extends `Listenable` and adds a single but very important property – a value. At this point, we have seen a mechanism for adding and removing listeners, as well as storing a value. What we haven't seen though is a mechanism that notifies the listeners that something has changed.

For that, we have `ChangeNotifier`, which is also an implementation of `Listenable`:

```
#1   class ChangeNotifier implements Listenable {
#2     int _count = 0;
#3     List<VoidCallback?> _listeners = [];
#4
#5     @protected
#6     bool get hasListeners => _count > 0;
#7
#8     @override
#9     void addListener(VoidCallback listener) {...}
#10
#11    @override
#12    void removeListener(VoidCallback listener) {...}
#13
#14    @mustCallSuper
#15    void dispose() {
#16      // ...
#17      _listeners = [];
#18      _count = 0;
```

```
#19    }
#20
#21   void notifyListeners() { ... }
#22 }
```

In the preceding code, we should take note of the following:

- On line #1, we can see that ChangeNotifier implements Listenable.

- On lines #2 and #3, we can see that there are internal fields that reference the count and the list of listeners.

- On lines #8 to #12, we can see that ChangeNotifier overrides the methods from Listenable. The code there is omitted because the details are irrelevant to our understanding. All we need to know is that inside those methods, manipulations occur with the _count and _listeners fields.

- On lines #14 to #19, we can see a new method called dispose. It removes all of the references to the listeners and sets the count to 0. We will return to this method again in the next section.

- The most interesting method for us right now is on line #21 –notifyListeners. While the implementation details don't matter, it is crucial to understand what this method does: it notifies its _listeners that there have been changes.

We've just taken a deep dive into the internals of Flutter. Now, let's shift our focus to a more practical perspective and explore how we can use these tools to manage the state.

Listening to changes via ValueNotifier

While ChangeNotifier provides us with a mechanism to notify listeners, it also requires us to handle the actual values ourselves. We will see how to do this and why it is useful in just a bit, but for now, let's look at an implementation of ChangeNotifier that's present in the Flutter framework – that is, ValueNotifier:

```
class ValueNotifier<T> extends ChangeNotifier implements
ValueListenable<T> {

ValueNotifier(this._value);

  @override
  T get value => _value;
  T _value;
  set value(T newValue) {
    if (_value == newValue) {
      return;
```

```
      }
      _value = newValue;
      notifyListeners();
    }
  }
```

The preceding class is very simple and adds only one thing to ChangeNotifier – a value of a generic type, T. This is a convenience class for cases when you only need to manage one value and don't want to create a full ChangeNotifier just for that. One more important thing here is that it implements ValueListenable – let's see why.

Let's return to our Candy Shop app. So far, we have been passing our Map<int, CartListItem> items as a parameter from MainPage to CartPage. In CartPage, we made a local copy of items to see changes instantly and used callbacks to propagate changes back to MainPage. However, this approach was quite cumbersome. To simplify things, we can refactor items to be a ValueNotifier class in MainPage, like this:

lib/main_page.dart

```
class _MainPageState extends State<MainPage> {
  // Before:
  // final Map<String, CartListItem> items = {};
  // After:
  ValueNotifier<Map<String, CartListItem>> items = ValueNotifier({});
}
```

In the preceding code, we wrapped our Map into a ValueNotifier class and gave it a default value of an empty map. We also need to refactor CartPage. Some code has been omitted so that we can focus on the relevant changes since we are only changing a single line:

lib/cart_page.dart

```
class CartPage extends StatefulWidget {
  // Before:
  // final List<CartListItem> items;
  // After:
  final ValueNotifier<Map<String, CartListItem>> items;
}
```

As a result of the preceding code, instead of a List value, CartPage accepts ValueNotifier. The best part is that there is a special widget in Flutter called ValueListenableBuilder that puts everything together. It accepts two parameters: valueListenable, which is of the ValueListenable type (the same type that's implemented by ValueNotifier), and a builder callback that returns an instance of a widget based on the value of ValueNotifier. This builder is called every time the underlying value in ValueNotifier changes. Now, let's refactor our CartPage so that it makes use of our ValueNotifier:

lib/cart_page.dart

```
class _CartPageState extends State<CartPage> {
  @override
  Widget build(BuildContext context) {
    return Scaffold(
      appBar: AppBar(
        title: const Text('Cart'),
      ),
```

Listen to ValueNotifier from the widget:

```
      body: ValueListenableBuilder(
        valueListenable: widget.items,
        builder: (context, items, _) {
          final values = items.values.toList();
          final totalPrice = items.values.fold<double>(0,
            (previous, element) => previous + element.product.
            price * element.quantity,
          );

          return Stack(
            children: [
              Padding(
                padding: const EdgeInsets.only(bottom: 60),
                child: ListView.builder(
                  padding: const EdgeInsets.symmetric(vertical: 16),
```

Use `values` from the builder here instead of `_items`:

```
itemCount: values.length,
itemBuilder: (context, index) {
  final item = values[index];
  return CartListItemView(
    item: item,
```

Invoke the widget callbacks without updating the local state:

```
          onRemoveFromCart: widget.onRemoveFromCart,
          onAddToCart: widget.onAddToCart,
        );
      },
    ),
  ),
  Positioned(...), //
],
    );
  }),
);
    }
  }
```

In the preceding code, we did a couple of things:

- We wrapped our body in `ValueListenableBuilder` and supplied it with `valueListenable`, which is the `ValueNotifier` parameter's `items` that we passed to the `CartPage` constructor.

- We eliminated all of the code related to maintaining a local copy of `items`! By using the `onAddToCart` and `onRemoveFromCart` widget callbacks, which update `ValueNotifier`, which is holding our cart items in `MainPage`, our `CartPage` will be notified and rebuilt with relevant data. This makes it super easy to use and we have removed much more code than we added.

Now, our data flow diagram looks slightly different:

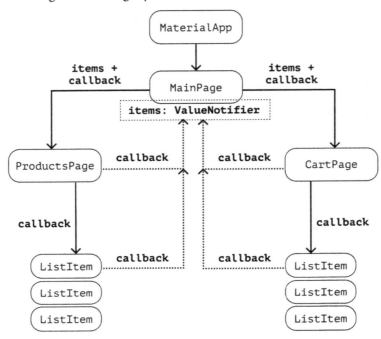

Figure 3.5 – The data flow when using ValueNotifier

First of all, the type of `items` is now `ValueNotifier`, but what's more important is that our `CartPage` doesn't store an `items` version of its own at all!

We refactored our items list to be observable by using `ValueNotifier`. However, there are other fields, such as `totalCount` and `totalPrice`, that still need to be calculated every time our list changes. This results in duplicated code, and potential efficiency problems as the list grows, and we're still passing a lot of callbacks back and forth, making our widgets tightly coupled. To solve this problem, we can go further and refactor using `ChangeNotifier`, which we learned about in the *Implementing an observable using the Listenable class* subsection earlier in this section.

Encapsulating state via ChangeNotifier

While `ValueNotifier` may be useful for simple single values, the reality is that even our cart information isn't that simple. We have a list of grouped cart items, as well as the total number of items and the total price. Rather than making each of these a `ValueNotifier` class, we can use one `ChangeNotifier`. This approach not only exposes multiple fields but also encapsulates all of the related business logic, removing it from `MainPage`. If we continue using the current pattern, `MainPage` will become a bloated monster of callbacks as our app grows. Instead, we can split the logic, unit test it via `ChangeNotifier`, and make our code less coupled.

So, first, let's create an instance of ChangeNotifier that will encapsulate all of the cart logic and call it CartNotifier:

lib/cart_notifier.dart

```
class CartNotifier extends ChangeNotifier {
   final Map<String, CartListItem> _items = {};
   double _totalPrice = 0;
   int _totalItems = 0;

   List<CartListItem> get items => _items.values.toList();
   double get totalPrice => _totalPrice;
   int get totalItems => _totalItems;
}
```

So far, everything is simple: we have internal fields for items, totalPrice, totalCount, and their public getters. Now, let's add some logic:

lib/cart_notifier.dart

```
class CartNotifier extends ChangeNotifier {
   // Fields here

   void addToCart(ProductListItem item) {
     // Exactly the same logic as we had in MainPage
     notifyListeners();
   }

   void removeFromCart(CartListItem item) {
     // Exactly the same logic as we had in MainPage
     notifyListeners();
   }
}
```

In the preceding code, we added two methods: addToCart and removeFromCart. The implementation is the same as in MainPage; nothing has changed. The only addition is a call to the notifyListeners method after all of the logic is executed. This will notify everyone who has subscribed to CartNotifier that there has been a change, in the same way we saw with ValueNotifier.

Now, there are a couple more things we need to do to make it work. First of all, we need to change the type of the parameter in `MainPage` and `CartPage`:

lib/main_page.dart

```
class _MainPageState extends State<MainPage> {
  CartNotifier cartNotifier = CartNotifier();
}

class CartPage extends StatefulWidget {
  final CartNotifier cartNotifier;
}
```

Now, our pages have a reference to `CartNotifier`. Next, we need to register the listeners. We will start with `MainPage`:

lib/main_page.dart

```
@override
  void initState() {
    super.initState();

    cartNotifier.addListener(() {
      setState(() {});
    });
  }

  @override
  void dispose() {
    cartNotifier.dispose();
    super.dispose();
  }
```

Our listener in the preceding code is quite simple: for now, we just rebuild the entire widget whenever something changes, and we achieve this through `setState`. While this may not be the most efficient practice since it re-renders the whole screen, instead of only the parts that have been updated, it suffices for this demo's sake. Another important consideration is that we must dispose of `cartNotifier` when the `MainPage` widget, which initially created `cartNotifier`, gets disposed of. The `dispose` method removes all listeners, ensuring that no memory is leaking and that no one is listening to a notifier that no longer exists.

We also need to update `CartPageState` so that it listens to the notifier:

lib/cart_page.dart

```
@override
void initState() {
  super.initState();
  widget.cartNotifier.addListener(_updateCart);
}

@override
void dispose() {
  widget.cartNotifier.removeListener(_updateCart);
  super.dispose();
}

void _updateCart() {
  setState(() {});
}
```

As evident from the preceding code, to add the listener, we can follow the same process as before. The `_updateCart` method simply calls `setState`. Since `CartPageState` did not create `cartNotifier` and instead received it as an argument in the constructor, we only need to remove the listener rather than dispose of the entire notifier.

And now comes the best part – we can remove the callbacks from `CartPage`:

lib/cart_page.dart

```
class _CartPageState extends State<CartPage> {
  // ...

  @override
  Widget build(BuildContext context) {
    // ... ListView.builder code here that builds CartListItemView
    return CartListItemView(
      item: item,
      // Calling methods on `widget.cartNotifier`
      // instead of callbacks
      onRemoveFromCart:
          widget.cartNotifier.removeFromCart,
      onAddToCart: (item) =>
          widget.cartNotifier.addToCart(item.product),
    );
```

```
    }
  }
```

As you can see in the preceding code, we are not only accessing items directly from CartNotifier but also calling its callbacks, thus removing them as fields from CartPage. We can go further and remove the arguments from CartListItemView and just pass the notifier itself, making it even cleaner. This way, we can refactor our entire app and decouple it from constructor callbacks.

However, up to this point, we have imperatively added and removed listeners in CartPage, which isn't super convenient and it's easy to forget to remove a listener. To address this, we can use ListenableBuilder, a convenience widget for ChangeNotifier. It works in the same way as ValueListenableBuilder, except that it accepts a generic Listenable, not a specific ValueListenable. It handles adding and removing listeners for us. It accepts a plain Listenable as a parameter, which our ChangeNotifier is.

We can use ListenableBuilder in CartPage like this:

lib/cart_page.dart

```
ListenableBuilder(
      listenable: widget.cartNotifier,
      builder: (context, _) {
        return Stack(
          children: [
            Padding(
              padding: const EdgeInsets.only(bottom: 60),
              child: ListView.builder(
                padding: const EdgeInsets.symmetric(vertical: 16),
                itemCount: widget.cartNotifier.items.length,
                itemBuilder: (context, index) {
                  final item = widget.cartNotifier.items[index];
                  return CartListItemView(
                    item: item,
                    cartNotifier: widget.cartNotifier
                  );
                },
              ),
            ),
            Positioned(...),
          ],
        );
      },
    )
```

In the preceding code, we removed the ability to register and unregister listeners from `cartNotifier`. Instead, we now listen to changes in `cartNotifier` directly via `ListenableBuilder`.

Let's take a look at the updated data flow diagram:

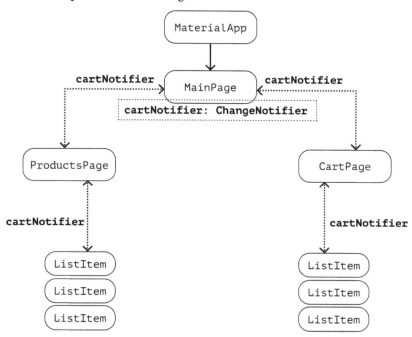

Figure 3.6 – The data flow when using ChangeNotifier

At this point, we have completely removed the business logic from `MainPage`. The page is now small and only maintains an instance of `CartNotifier`. Our widgets now declaratively observe changes in `CartNotifier` and update reactively when changes occur. We no longer pass callbacks such as `onRemove` and `onAdd` in constructors. However, we still pass around the `cartNotifier` instance in constructors, so the coupling is not as tight but still present.

The last problem we face is passing around `cartNotifier`. The main page needs to know everything, and this makes it harder to change the entry point if we need to pass in many dependencies. Fortunately, Flutter provides a solution for this as well.

Passing around dependencies via InheritedWidget

The Flutter mechanism for passing around dependencies through the tree is called `InheritedWidget`. You have certainly used it in your Flutter apps, even if you haven't written one explicitly. Let's take a look at what `InheritedWidget` is and how it can help us on our state management journey.

What is InheritedWidget?

As you know, in Flutter, *everything is a widget*. So far, we have discussed various UI-related building widgets, such as Stateless, Stateful, and Render, as well as their descendants. However, because these widgets are organized in a tree data structure, it is possible to perform various manipulations with it, such as a tree traversal. This capability is useful when we need to not only render static UI but also pass around shared data.

The Flutter framework includes a widget specifically for this purpose: InheritedWidget. It is the last of the fundamental Flutter widgets. If we examine the framework.dart class (https://github.com/flutter/flutter/blob/master/packages/flutter/lib/src/widgets/framework.dart) and search for an abstract class, we will find only 25 instances in the entire file (as of Flutter 3.10). All of these are in some way related to Stateless, Stateful, Render, or Inherited.

So, let's take a look at the source code of InheritedWidget:

```
abstract class InheritedWidget extends ProxyWidget {
    const InheritedWidget({ super.key, required super.child });

    @override
    InheritedElement createElement() => InheritedElement(this);

    @protected
    bool updateShouldNotify(covariant InheritedWidget oldWidget);
}
```

In the preceding code, take note of the following:

- InheritedWidget extends ProxyWidget. Here, ProxyWidget is just an abstract class that extends Widget and has a single parameter – the child widget. It is used as a base widget.

- Then, we pass the child parameter in the constructor so that InheritedWidget will be a wrapper around some other widget.

- Then, we can see the already familiar createElement method, which creates InheritedElement. We won't stop here since the main logic behind it is the same as with the other widget elements.

- Now, we come to the most interesting part – the updateShouldNotify method, which returns a bool value and accepts an oldWidget value of the same type as a parameter. In the override of this method, we determine whether there are any differences that we care about in the old instance of the widget and the new one. If there are (meaning we return true), those changes are then propagated to everyone who inherits from this widget.

But how can we inherit from this widget and what kind of data may we possibly want to pass around? We don't need to go far to find examples – you have probably already used them in your app. Let's take a deeper look into these `.of(context)` methods.

Understanding the .of(context) pattern

Some of the most common actions that are performed in apps include theming, screen navigation, and building UI based on screen size. To accomplish these tasks in Flutter, we can use `Theme.of(context)`, `Navigator.of(context)`, and `MediaQuery.of(context)`. As a Flutter developer, you can often use these widgets without even thinking about why they work. We rarely explicitly specify `MediaQuery` or `Navigator` in our widget tree, yet we can still access them and they work perfectly fine. How is this possible?

The answer is that Flutter does this for us. Most of the time, the root of our widget tree is either `MaterialApp` or `CupertinoApp`, which both return `WidgetsApp` under the hood. `WidgetsApp` wraps our widgets in `MediaQuery`, `Navigator`, or `Router` widgets. At different levels of nesting and abstractions, all of these widgets extend `InheritedWidget`. `InheritedWidget` is a special type of widget that allows us to access it from wherever we are in the tree, so long as the instance of `InheritedWidget` can be found in the tree.

To better understand this concept, let's look at an actual example. You can find the full source code here: `https://github.com/flutter/flutter/blob/master/packages/flutter/lib/src/widgets/media_query.dart`. Let's examine the part of the source code that allows us to access `MediaQuery`:

```
class MediaQuery extends InheritedWidget {
  final MediaQueryData data;

  static MediaQueryData of(BuildContext context) {
    return context
        .dependOnInheritedWidgetOfExactType<MediaQuery>()!.data;
  }
}
```

Let's take a look at the preceding code:

- The first thing we notice is a field called `data` of the `MediaQueryData` type.

- The second thing we notice is that the `.of` method returns `MediaQueryData`.

- Finally, to obtain this `data` from the `.of` method, we must call the `dependOnInheritedWidgetOfExactType<MediaQuery>` method on the `context` parameter. We will discuss this method and `BuildContext` in more detail shortly. For now, we should remember two things:

- When we call this method, Flutter searches the widget tree from the current widget to the root for a widget of this type – in this case, MediaQuery. If it finds one, it returns its data property. If the widget does not exist in the widget tree, Flutter throws an error.

- In addition to locating and returning InheritedWidget (in this case, MediaQuery), this method adds the widget that calls it to Flutter's internal list of dependent widgets. Whenever something in the inherited widget changes (meaning updateShouldNotify returns true), all subscribed widgets are rebuilt.

Let's remember why we're looking at InheritedWidget. So far, we've been passing around CartNotifier as a parameter via the constructor. This tightly couples our code and makes it less flexible. In Flutter, we have an alternative approach: provide this notifier via a widget higher up in the tree, and then read it from context wherever we need it, similar to how we use MediaQuery. So, let's implement this approach.

Creating the CartProvider class

Let's create a class called CartProvider that will provide CartNotifier in the widget tree. To do that, we need to extend an InheritedWidget class:

lib/cart_notifier_provider.dart

```
#1 class CartProvider extends InheritedWidget {
#2   final CartNotifier cartNotifier;
#3
#4   const CartProvider({
#5     super.key,
#6     required this.cartNotifier,
#7     required Widget child,
#8   }) : super(child: child);
#9
#10  static CartNotifier of(BuildContext context) {
#11    return context
#12        .dependOnInheritedWidgetOfExactType<CartProvider>()!
#13        .cartNotifier;
#14  }
#15
#16  @override
#17  bool updateShouldNotify(CartProvider oldWidget) {
#18    return cartNotifier != oldWidget.cartNotifier;
#19  }
#20 }
```

Here's what's going on in the preceding code:

- On line #1, we extend `InheritedWidget`.

- On line #2, we create an internal field that we want to provide in our widget tree – the instance of `CartNotifier` – and call it `cartNotifier`.

- On lines #4 to #8, we create a constructor, in which we have the standard `key` and `child` params, which are required by `InheritedWidget`, and `cartNotifier`, which is required by `CartProvider`.

- On lines #10 to #12, we follow the same `.of(context)` pattern that's omnipresent in Flutter, and in the same way we have seen with `MediaQuery`, we return `CartNotifier`.

- On lines #15 to #17, we override `updateShouldNotify` and return `true` in case the instances of `cartNotifier` are not the same.

The next thing we need to do is to provide an instance of `CartProvider` in our widget tree. Because we will need `CartProvider` on almost every page of our app, it makes sense to have it at the very root. So, let's wrap `MaterialApp` in `CartProvider`:

lib/main_page.dart

```
void main() {
  runApp(
    CartProvider(
      cartNotifier: CartNotifier(),
      child: MaterialApp(
        title: 'Candy shop',
        theme: ThemeData(
          primarySwatch: Colors.lime,
        ),
        home: const MainPage(),
      ),
    ),
  );
}
```

Now, the root widget of our app is `CartProvider` and we can access it anywhere in our widget tree. Let's refactor our code so that it does that.

For example, in `ProductListItemView`, instead of passing any callback as params to constructors, we can just use `CartProvider` to find `CartNotifier`, like this:

lib/product_list_item_view.dart

```
class ProductListItemView extends StatelessWidget {
  final ProductListItem item;

  const ProductListItemView({
    Key? key,
    required this.item,
  }) : super(key: key);

  @override
  Widget build(BuildContext context) {
    final cartNotifier = CartProvider.of(context);
    // read & submit data to/from cartNotifier
  }
```

As a result of the preceding code, we can create `ProductListItemView` from anywhere, without worrying about providing callbacks or how deeply nested those callbacks are. This way, we avoid what is known in programming as *callback hell*, as well as nicely encapsulate the business logic away from our widgets. We can refactor all of the other widgets that use `CartNotifier` in the same way we did in `ProductListItemView`.

Let's look at the final version of the data flow diagram:

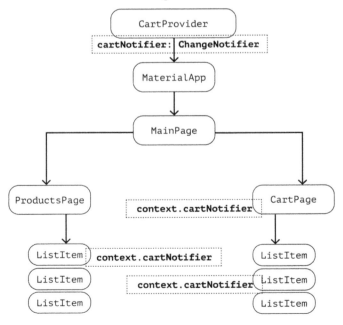

Figure 3.7 – The data flow when using CartProvider

At this point, we don't pass any dependencies as the constructor parameters and we can fetch `cartNotifier` via `context` in any part of the subtree below `CartProvider` that we need!

We have solved a lot of problems so far! Let's sum it up by remembering which problem was solved by which tool:

Problem	Solution
Update UI based on local state changes in a single widget	`setState` method in `StatefulWidget` class
Update UI based on shared state changes in several widgets	*Lifting state up* pattern
Remove the *callback hell* issue introduced by the *Lifting state up* pattern	Observe changes via `ValueNotifier` and `ValueListenableBuilder` classes
Observe and change more than a single value	`ChangeNotifier` class and listeners
Remove potential memory leaks problem introduced by `ChangeNotifier` listeners	`ChangeNotifier` and `ListenableBuilder` classes
Remove tight coupling of `ChangeNotifier` to `MainPage` for more flexible dependency management	`InheritedWidget` class via generic `.of(context)` pattern
Avoid errors being thrown due to widget not being present in the widget tree	`InheritedWidget` class via `.maybeOf(context)` pattern
Avoid redundant widget rebuilds caused by changes in unused parameters	`InheritedWidget` class via specific `.of(context)` pattern

Table 3.2 – Summary of the problems solved and their respective tools

Before we wrap up this chapter, there are a couple more things you need to know about `InheritedWidget` and its relationship with `BuildContext` to write more stable and less bug-prone code.

Interacting with BuildContext in the right place, at the right time

So far, we have indirectly mentioned `BuildContext` here and there, but we have never really given it the attention it deserves. And it deserves a lot because `BuildContext` is a crucial and fundamental concept of the Flutter framework. So, let's take a closer look at what it is, how to use it, and what its role is in state management. You can find the full source code here: https://github.com/flutter/flutter/blob/master/packages/flutter/lib/src/widgets/framework.dart.

What is BuildContext?

There are two main cases where we interact with `BuildContext`. First, it is used in the `build` method of a widget (`Widget build(BuildContext context)`, where it is passed as a parameter. We mainly use the context from this method to access inherited widgets such as `MediaQuery`. Essentially, we need the context to locate widgets in the widget tree. Now, let's take a closer look at the source code for `BuildContext`:

```
abstract class BuildContext {
  // Some code omitted for demo purposes
  T? findAncestorWidgetOfExactType<T extends Widget>();

  T? findAncestorStateOfType<T extends State>();

  T? findRootAncestorStateOfType<T extends State>();

  T? findAncestorRenderObjectOfType<T extends RenderObject>();

  InheritedWidget dependOnInheritedElement(InheritedElement ancestor,
{ Object aspect });

  T? dependOnInheritedWidgetOfExactType<T extends InheritedWidget>({
  Object? aspect });
}
```

The first thing we can see is that `BuildContext` is an abstract class, meaning it defines an interface that someone has to implement. It also has several methods related to finding ancestors of different types, such as widgets, state, and render objects. This confirms that the context is used to locate various objects in our widget tree. The next set of methods is similar in the sense that they find some kind of object, but different in the sense that they also depend on that object. We previously used the `dependOnInheritedWidgetOfExactType` method when working with inherited widgets. So, essentially, `BuildContext` is just that – an object locator. The most interesting part is who implements the `BuildContext` interface because we're already familiar with this concept. The last time that I showed you this concept's source code in the *Unveiling the Element class* section in *Chapter 1*, I omitted one crucial part on purpose. Let's see if you can spot it now:

```
abstract class Element implements BuildContext {}
```

Yes, `BuildContext` is none other than our `Element`. As we learned previously, widgets are just configuration objects for the actual tree, which is not a widget tree, but an element tree. Also, every widget is backed up by an element. The code documentation (`https://api.flutter.dev/ flutter/widgets/BuildContext-class.html`) states this:

"The [BuildContext] interface is used to discourage direct manipulation of [Element] objects."

So, technically, Flutter could have used Element directly, but it's better coding practice to separate responsibilities and make the APIs as tight as possible. Now, let's review some problems that can be caused by the misuse of BuildContext.

Context does not contain the widget

Naturally, if something can be looked up, there is always a chance that it won't be found. This is especially true when looking up widgets, such as through the familiar .of(context) pattern. You may have already encountered the most notorious of these problems:

Scaffold.of() called with a context that does not contain a Scaffold. No Scaffold ancestor could be found starting from the context that was passed to Scaffold.of(). This usually happens when the context provided is from the same StatefulWidget as that whose build function actually creates the Scaffold widget being sought.

The error message itself often provides reasons why the error may occur, with possible solutions included. The Scaffold widget is a notable example, but in reality, this can happen with any kind of widget, particularly when working with inherited widgets for state management. It is crucial to remember to provide the intended widget in the correct location in the tree. In some cases, such as with the Theme widget, a default widget is returned if none can be found in the context. However, when working with our own inherited widgets, such as CartProvider, a null pointer exception may be thrown if the desired widget is not found. It is best practice to throw a more specific error message for easier debugging. Therefore, we could refactor our .of method in the following way:

lib/cart_notifier_provider.dart

```
static CartNotifier of(BuildContext context) {
    final provider =
      context.dependOnInheritedWidgetOfExactType<CartProvider>();

    if (provider == null) {
      throw Exception('No CartProvider found in context');
    }

    return provider.cartNotifier;
}
```

Here, we added a null check. If the provider is null, we throw a specific error. This way, the user of the widget will at least have an idea of what the problem is.

Exploring beyond `.of(context)` **pattern**

As we have just learned, if we try to access a widget via `.of(context)` pattern and the requested widget is not found in the widget tree, the framework will throw an error. While in some cases this may be the desired behaviour as it indicates erroneous code, sometime we may be aware that there is a chance that the widget won't be present and we know how to hande that. For that, we have an alternative pattern – `.maybeOf(context)` – which, instead of throwing an exception, will return a `null` value if the widget is not present in the widget tree. You can then handle the `null` value accordingly. Moreover, accessing `MediaQuery` via the generic `.of(context)` method will trigger rebuilds if *any* of the property changes, be it padding, view insets, screen size, and so on. Often, we only care about specific values, so we can subscribe to specific changes by using specific methods, such as `.paddingOf(context)`, `.sizeOf(context)`, and so on.

Context contains the widget, but not the one you expect

On the opposite side, there may be a situation where not just one widget of the exact type is present, but many. When accessing that widget, you may expect to receive one value, but instead receive another. The problem may be that there is another widget of the same type between the calling widget and the widget you anticipate. Because the ancestor lookup starts from the calling widget and returns the closest ancestor it can find, you may receive another instance. In this case, you need to examine the widget tree and locate the unexpected widget. For example, this can occur if you use `SafeArea` in your widget tree. `SafeArea` overrides `MediaQuery` and manages its insets. Therefore, if you attempt to access `MediaQuery` lower in the tree, you may receive unexpected results. This can also occur with providers, so make sure you do not provide the same provider more than once; otherwise, your state may be inconsistent.

Context accessed too early!

`BuildContext` is closely related to the life cycle of the widget. Although the `StatelessWidget` life cycle may not cause many problems, the life cycle of `StatefulWidget`, or rather its `State`, is something to keep in mind when accessing inherited widgets.

Returning to `dependOnInheritedWidgetOfExactType`, recall that it locates the widget in the tree and adds the calling widget to an internal list of dependents. If the widget changes, it can notify the dependents via the `didChangeDependencies` method of `State` or by rebuilding `StatelessWidget`. It is important to note that widgets accessed via `dependOnInheritedWidgetOfExactType` should only be accessed from methods that can be called several times during the life cycle. For `State`, those methods are `build` and `didChangeDependencies`. It is safe to access those widgets in those methods.

On the other hand, accessing a widget via this method in the `initState` method of `State` will result in an error. This is because `initState` is called only once per life cycle of `State`, meaning that if you try to subscribe your widget inside of this method, it can never receive updates. Flutter prohibits this and will throw an error. So, make sure you are not accessing your `InheritedWidget` too early.

Sometimes, it is useful to look at what's happening inside the `.of(context)` method. Also, this only applies to widgets that can be rebuilt (in other words, use the `dependOnInheritedWidgetOfExactType` method of `BuildContext`). Inherited widgets that are accessed only once and are not dependent on can be accessed in `initState` too.

For example, let's try to access `MediaQuery` in `initState`:

```
@override
  void initState() {
    super.initState();
    final size = MediaQuery.of(context).size;
  }
```

We will see the following runtime error:

dependOnInheritedWidgetOfExactType<MediaQuery>() or dependOnInheritedElement() was called before _CartPageState.initState() completed. When an inherited widget changes, for example if the value of Theme.of() changes, its dependent widgets are rebuilt. If the dependent widget's reference to the inherited widget is in a constructor or an initState() method, then the rebuilt dependent widget will not reflect the changes in the inherited widget.

This error message confirms what we have just learned: it is prohibited to call `dependOnInheritedWidgetOfExactType` in the `initState` method.

Context accessed too late!

Another frequent issue that's caused by `BuildContext` is when it's captured in async gaps. As we have seen, widgets have life cycles, especially `StatefulWidget`. There may be a situation where `context` is accessed too late. This can happen if you capture it in an `async` gap. We will discuss `Future` and `async`/`await` in more detail in *Chapter 9*, but for now, remember that if you use `context` after executing something with an `await` syntax, there is a chance that your `State` was already disposed of via its `dispose` method, or your `StatelessWidget` was removed from the tree, meaning the context has become `unmounted`. If you have saved a reference to your context and then accessed it when it has become `unmounted`, you will get an error. You can only use `context` when it's `mounted`, meaning that the widget is present and active in the widget tree. To avoid this error, first, enable the `use_build_context_synchronously` lint rule to highlight places like this. To fix it, check if `context` is mounted before accessing it after an async gap, like this: `if` (context.mounted) { // do something with context }. An example of this scenario is shown here:

```
GestureDetector(
  onTap: () async {
    // The `await` keyword here
    await Future.delayed(const Duration(seconds: 10));
    // Uncomment next line to fix the issue
    // if (context.mounted)

    // `context` is captured here
    ScaffoldMessenger.of(context).showSnackBar(
      SnackBar(
        content: Text('I have been tapped'),
      ),
    );
  },
);
```

In the preceding code, when we use await, all the code *after* it is saved by Flutter so that it can be executed after the Future is completed, hence the name *async gap*. We also save the reference to context, which could have been unmounted during the 10 seconds. Using it unconditionally for widget lookup can result in an error. To avoid this, we need to uncomment the line with the context. mounted check, which only executes the lookup code if the context is still valid.

With that, we know how to safely work with BuildContext, which means that we can also build more reliable and bug-free state management solutions.

Summary

We learned a lot in this chapter! We observed the various types of states that an app can have and why these states should be managed. We also learned what tools and mechanisms Flutter has out of the box for managing state: starting from the basic setState of StatefulWidget and taking a deeper dive into Listenable, ValueListenable, ValueNotifier, ChangeNotifier, and the widgets that go with them – ValueListenableBuilder and ListenableBuilder. We then explored how to connect those tools with another staple Flutter widget, InheritedWidget, and understood how it works. Finally, we reviewed the role of BuildContext in state management and Flutter, as well as how to avoid the most common errors related to its misuse.

An important thing to note here is that to do proper state management, you already have everything in vanilla Flutter and it works perfectly fine. We have built a clean, decoupled, and maintainable solution by using only the tools that are available out of the box. All of the third-party state management solutions are based on top of this foundation and only add to it. We have also seen what problems the vanilla state management introduces, such as a lot of boilerplate code and a lot of gotchas that we need to remember.

In *Chapter 4*, we will review what state management design patterns exist in software development in general, which of them suit mobile applications the best, and get hands-on practice with some of the most popular state management libraries from the Flutterverse.

Get this book's PDF version and more

Scan the QR code (or go to `packtpub.com/unlock`). Search for this book by name, confirm the edition, and then follow the steps on the page.

Note: Keep your invoice handy. Purchases made directly from Packt don't require an invoice.

4

State Management Patterns and Their Implementations

There are two notions in state management that are often confused but are quite different: patterns and tools. While a lot of developers argue about the tools, the real thing that you should consider is the pattern, and you should select your tools based on the pattern that suits you the best.

But what do we mean when we say tools or patterns?

Patterns are like guidelines or instructions on how to solve common problems. They're generally independent of the frameworks and programming languages and operate at a higher level of abstraction. Meanwhile, tools are specific and often opinionated implementations of the patterns. For example, MVI is a pattern, but the `flutter_bloc` library is a tool that implements it.

In the previous chapter, we learned about the importance of proper state management for building Flutter apps. We created a solution by using only the tooling available in the Flutter SDK. We did that based on a simple example. In this chapter, we will dive even deeper into state management techniques, as well as handle more complicated use cases.

In this chapter, we will explore various state management patterns, such as MVVM, MVI, and BLoC, the ideas behind them, their pros and cons, and their practical implementations. We will continue working on our Candy Store app from the previous chapter as we will be refactoring it step by step to observe various patterns.

By the end of this chapter, you will understand the theoretical and practical differences between popular state management patterns, learned how to analyze patterns to select which one suits your use case the best, and understand the hands-on approach to implementing those patterns both by using the vanilla Flutter SDK and the `flutter_bloc` library (https://bloclibrary.dev/#/).

In this chapter, we're going to cover the following main topics:

- Diving into the world of MVX patterns

- Defining criteria for the state management pattern

- Embracing data binding with MVVM

- Following the MVVM and MVI patterns with `flutter_bloc`

- Implementing the Segmented State Pattern with `DelayedResult`

- Avoiding bugs related to state equality and immutability

Technical requirements

To proceed with this chapter, you will need the following:

- The code from the previous chapter, which can be found here: `https://github.com/ PacktPublishing/Flutter-Design-Patterns-and-Best-Practices/tree/ master/CH03/final/candy_store`

- The code required for this chapter:

  - Start of this chapter: `https://github.com/PacktPublishing/Flutter- Design-Patterns-and-Best-Practices/tree/master/CH04/initial/ candy_store`

  - End of this chapter: `https://github.com/PacktPublishing/Flutter- Design-Patterns-and-Best-Practices/tree/master/CH04/final/ candy_store.`

  You can review the step-by-step refactoring in the commit history of this chapter's folder.

You will also need to add the `https://pub.dev/packages/flutter_bloc` and `https:// pub.dev/packages/equatable` libraries to the Candy Store app.

Diving into the world of MVX patterns

To understand what technique or pattern to select, we need to understand why they exist and what problems they solve. Once you start researching the topic of state management patterns and software design patterns in general, you're bound to bump into abbreviations such as MVC, MVVM, MVI, MVP, MVU, and others. Despite the variety of the letters in those abbreviations, there are a couple that always persist: *M* and *V*. Those letters always stand for two important yet opposite notions: **Model** and **View**. The letters connecting them are intermediaries between those two: *C* stands for **Controller** in the *MVC*, *VM* stands for **ViewModel**, *I* stands for **Intent** in *MVI*, and so on. It all started with MVC back in the late 1970s (`https://en.wikipedia.org/wiki/ Model%E2%80%93view%E2%80%93controller`), but over the years, the MVC pattern has been transformed by many interpretations and today it is hard to reason about how exactly this pattern is supposed to work as there are many correct answers. We will dive deeper into those MV implementations that make sense in Flutter, but first, let's observe the key players that stay the same in all of these patterns – the **Model** (**M**) and the **View** (**V**).

What is the View?

The *View* is the representation of the user's perspective. This is what they see on the screen when the data from the *Model* is converted into something that can be digested by the user, usually a GUI. In Flutter, the widgets represent the *View*. We learned how to work efficiently with this layer in *Part 1* of this book.

What is the Model?

The *Model* is the representation of the data and its operations: how to fetch data, from where to fetch it, how to store it, what its format is, and so on. It handles all of the "invisible" operations that go under the hood and are hidden from the user. From a practical point of view, this could be an API client, a database client, or any other source of data that is independent of its visual representation. The word *Model* itself can be confusing, so to distinguish it from any other meaning of the word "model," we will be using the capitalized, italicized version when referring to the *Model* notion from the MVX patterns. We will learn how to efficiently work with this layer in *Part 3* of this book.

How are they connected?

This chapter is dedicated to answering the question "How are the *Model* and the *View* connected?" because there isn't just a single and definite option. Up to this point, we have mentioned MVX, where M is the *Model* and V is the *View*, yet X is just a placeholder for many possibilities:

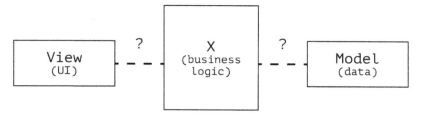

Figure 4.1 – How are the View and Model connected?

Moreover, the answer consists of not only answering what the **X** stands for but also how it facilitates communication between the *Model* and the *View*. This is where things can start to get a little bit confusing, especially when it comes to terminology. The goal of the component currently called X is to transform the data coming from the *Model* layer into the UI of the *View* layer (and the other way around), hence there needs to be some logic applied. Parts of this logic can be called presentation logic, other parts - application logic or business logic. You may come across pattern definitions that describe business logic as the concern of exclusively *Model* layer, others less strict about it. In reality, those boundaries are often not that definitive. As we learn about more patterns and continue building our application, we will see how this logic can shift from one layer to another. To be consistent, we will call it business logic.

We started working with this layer in *Chapter 3* and will continue to do so in *Chapter 4*. But before we review our options, we need to define the criteria for their assessment.

Defining criteria for the state management pattern

When selecting the most suitable pattern, there can be many criteria and it may vary from person to person. There will never be a situation when the whole world agrees on one specific pattern. Nevertheless, here are some important points that I believe hold a general appeal:

- The business logic should be independent of the UI logic and should be clearly separated.
- The pattern should be used consistently.
- At the same time, the pattern should be scalable and flexible.
- It should be comprehensible to the team. This criteria may be often overlooked as non-important, but in the end, the software is developed by people and the more effectively they can work with the pattern, the more productive the output will be.
- The pattern should complement the framework it is being used in, not work against it.

Some of the implementations of the MVC pattern violate these criteria:

- Specifically in that implementation, when the *View* knows of the *Model*, the coupling is unnecessarily tight. It is harder to swap one *Model* for the other without disturbing the *View*.
- In many implementations, the *Controller* can imperatively get hold of the *View* (in Flutter it could be access to `BuildContext`) and update its properties based on the changes in the *Model*. While this may work well in imperative style frameworks such as native Android before JetpackCompose and native iOS before SwiftUI, the reactive and declarative design of the Flutter framework makes strictly implementing this pattern impossible without going against the framework. Now, let's tap into what patterns work well with reactive paradigms and why.

We have already discussed the declarative nature of Flutter many times. It can be expressed in a more mathematical formula. This is the figure that you can find in the official documentation (13.08.23, `https://docs.flutter.dev/data-and-backend/state-mgmt/declarative`):

$$UI = f(state)$$

layout of   build   application state
the screen   methods

Figure 4.2 – UI equals the function of the state

This formula states that the UI always represents the state. We don't modify parts of the UI – we rebuild the whole UI based on the changes in the state. In the MVX terminology, by UI, we assume the *View*. Next, we're going to dive into the patterns that allow us to abstract away the details of state manipulation and just supply the result to the *View*.

Embracing data binding with MVVM in Flutter

One of the evolutions of the MVC pattern is MVVM, which stands for **Model-View-ViewModel**. It has the already familiar *Model* and *View*. But instead of a *Controller*, it has another mediator – the *ViewModel*. Its relationship with the *View* is heavily based on a concept called **data binding**.

Data binding is a technique that allows us to bind data between two notions: one that provides the data and one that consumes the data, in a way that if the data in the provider changes, it is automatically displayed in the consumer data. Let's take a look at a diagram that highlights this communication:

Figure 4.3 – Data flow among the MVVM components

The declarative and reactive design of the Flutter framework has this behavior out of the box. Because we declaratively build our UI as a reaction to state changes, there is no need for us to manually, imperatively update our *View*. For example, in a vanilla `StatefulWidget`, we build our UI based on the internal fields of `State` and when something changes, we flush the changes by calling `setState`. The widget tree then automatically rebuilds.

With some little tweaks, we can say that our current implementation from *Chapter 3* already resembles the MVVM. So, let's recap:

1. We have the *View*, which is the widget. It doesn't have any business logic in it. Instead, it takes values to render from `ListenableBuilder`, which emits them. It intercepts user interactions and passes them on to be handled by `CartNotifier`.

2. In this case, `CartNotifier` is the *ViewModel*. It doesn't know anything about the *View*; it handles business logic and saves it in the internal state. When anything in that state changes, it propagates those changes by notifying its listeners via `notifyListeners`. Anyone who has subscribed to these changes can react (observe) and update them accordingly.

3. There is no *Model* in our equation. All of the *Model* logic is stored in `CartNotifier` itself. So, the first step would be to extract the data handling into a *Model*.

Now, we're going to refactor our existing implementation so that it closely follows the MVVM architecture so that we can adhere to the criteria for a good state management solution, as we defined earlier. Before we dive into coding, let's quickly overview how MVVM complies with our criteria:

Criteria	MVVM
The business logic should be independent of the UI logic and should be clearly separated.	In MVVM, the *View* is responsible for the UI, and the *ViewModel* and *Model* for the business logic. The *ViewModel* isn't aware of the *View*, the logic is clearly separated.
The pattern should be used consistently.	It's possible to use MVVM consistently throughout the whole app.
The pattern should be scalable, flexible and testable.	Due to clearly defined relationships between actors and their responsibilities, MVVM scales very well.
The pattern should be comprehensible to the team.	As mentioned previously, because the pattern is clearly defined, the team can always refer to the definition as the source of truth.
The pattern should complement the framework it is being used in, not work against it.	Flutter's declarative UI-building approach embraces the MVVM data binding approach naturally.

Table 4.1 – MVVM adherence to good state management criteria

Similarly, you can analyze any potential approach to understand whether it suits your use case and criteria or not. Now that we know that MVVM is a good candidate for our solution, let's start implementing it. We will do so by introducing the *Model*. Before we dive into the code, let's look at a diagram of the components and their relationships. You can also refer to this diagram if you feel lost during our refactoring:

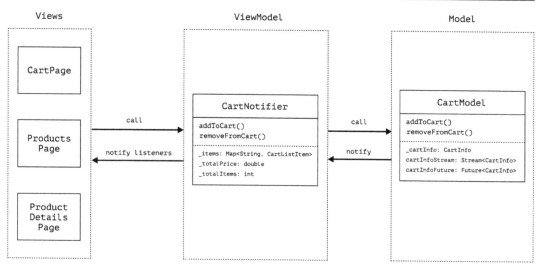

Figure 4.4 – Data flow among the Candy Store app components

By comparing the MVVM data flow diagram and this diagram, we can already see how close we are to following this pattern. Now, let's start coding.

Extracting data logic into the Model

We will learn how to best structure our *Model* layer in *Chapters 6*, *7*, and *8*, so for now, we will do the bare minimum so that we can focus on the pattern at hand. The idea is to introduce the *Model* layer by extracting all of the actions related to manipulating data from `CartNotifier` into a separate class called `CartModel`.

Following the encapsulation principle of OOP, let's combine the fields related to cart information into a single class called `CartInfo`, which will be responsible for describing the current state of the cart. It will look like this:

lib/cart_info.dart

```
class CartInfo {
  Map<String, CartListItem> items;
  double totalPrice;
  int totalItems;

  CartInfo({
    required this.items,
    required this.totalPrice,
    required this.totalItems,
```

```
    });
  }
```

Next, let's extract the addition and removal logic into a new class that will represent our *Model* and call it `CartModel`:

lib/cart_model.dart

```
#1   class CartModel {
#2     CartInfo _cartInfo = CartInfo(
#3       items: {},
#4       totalPrice: 0,
#5       totalItems: 0,
#6     );
#7
#8     CartInfo get cartInfo => _cartInfo;
#9
#10    void addToCart(ProductListItem item) {
#11      // Same logic as before
#12      // notifyListeners();
#13    }
#14
#15    void removeFromCart(CartListItem item) {
#16      // Same logic as before
#17      // notifyListeners();
#18    }
#19  }
```

Let's review what we have done line by line:

1. On lines 2 to 8, we saved a local field called `_cartInfo` that will store the information about the current state of the cart. We also defined a getter for it.

2. On lines 10 to 18, we kept the logic the same as we did previously, so we will omit it for brevity. However, one interesting thing you can see on lines 12 and 17 is that we have commented out `notifyListeners`. We did this because this method is specific to `ChangeNotifier`, while our model is not.

However, the return type of our methods is still `void` and we need to let the caller of these methods know that changes were made to the underlying data. How can we do this without `ChangeNotifier`? For that, we have the Streams API from Dart.

Emitting data via the Streams API

In *Chapter 3*, we discussed the observable pattern: we have some source of data that can emit changes (observable) and we have someone who listens to these data changes (the observer). Recall that Flutter has an implementation of this pattern called `Listenable` (of which `ValueNotifier` and `ChangeNotifier` are implementations). Well, pure Dart also has an implementation of this pattern and it's called `Stream`. You can create a `Stream` implementation of data, you can push new data to this `Stream`, and you can subscribe to this `Stream` to get updates on that data. Let's see it in practice:

lib/cart_model.dart

```
#1    class CartModel {
#2      CartInfo _cartInfo = {...}
#3
#4      CartInfo get cartInfo => _cartInfo;
#5
#6      final StreamController<CartInfo> _cartInfoController =
#7          StreamController<CartInfo>();
#8      Stream<CartInfo> get cartInfoStream => _cartInfoController.stream;
#9      Future<CartInfo> get cartInfoFuture async => _cartInfo;
#10     void addToCart(ProductListItem item) {
#11       // Same logic as before
#12       _cartInfoController.add(_cartInfo);
#13     }
#14
#15     void removeFromCart(CartListItem item) {
#16       // Same logic as before
#17       _cartInfoController.add(_cartInfo);
#18     }
#19
#20     void dispose() {
#21       _cartInfoController.close();
#22     }
#23   }
```

There are a couple of interesting things going on here:

1. On lines 6 to 8, we have created `Stream` and `StreamController`. Here, `Stream` is what stores the data and `StreamController` is what helps us manage how new data, as well as subscriptions, is emitted.

2. On lines 12 and 17, we used the add method of StreamController to add new data to our stream. What we passed there is _cartInfo. We modified it in the implementations of the addToCart and removeFromCart methods.

3. To notify our listeners that there will be no more events, we needed to call the close method of StreamController. We encapsulated this in the dispose method of CartModel so that we don't leak the details of the implementation to the outside.

4. On line 9, we created a Future that returns a single instance of CartInfo. This can be used for single reads when we don't need to listen to the updates.

Now, to align with the pattern terminology, let's rename CartNotifier to CartViewModel. We will also rename the file from cart_notifier.dart to cart_view_model.dart. Accordingly, we will rename CartProvider to CartViewModelProvider, along with its filename. So, what's left in CartViewModel and what does it do? Let's take a look inside!

lib/cart_view_model.dart (renamed from cart_notifier.dart)

```
class CartViewModel extends ChangeNotifier {
  final CartModel _cartModel = CartModel();

  CartViewModel() {
    _cartModel.cartInfoStream.listen((cartInfo) {
      _items.clear();
      _totalItems = cartInfo.totalItems;
      _totalPrice = cartInfo.totalPrice;
      cartInfo.items.forEach((key, value) {
        _items[key] = value;
      });
      notifyListeners();
    });
  }

  final Map<String, CartListItem> _items = {};
  double _totalPrice = 0;
  int _totalItems = 0;

  List<CartListItem> get items => _items.values.toList();

  double get totalPrice => _totalPrice;

  int get totalItems => _totalItems;

  void addToCart(ProductListItem item) {
    _cartModel.addToCart(item);
```

```
  }

  void removeFromCart(CartListItem item) {
    _cartModel.removeFromCart(item);
  }

  @override
  void dispose() {
    super.dispose();
    _cartModel.dispose();
  }
}
```

There are a couple of things we're doing here now:

1. We subscribe to the stream of `CartInfo` from `CartModel` in the constructor of `CartNotifier`. Now, every time there is a change in `cartInfoStream`, we will update our local fields and notify listeners as before – via the `notifyListeners` method.

2. In our `addToCart` and `removeFromCart` methods, we now have no logic and are just calling the methods of `CartModel`.

3. We have also renamed `CartNotifier` to `CartViewModel` so that it's easier to follow the terminology of the pattern and stay consistent.

Let's recap what we have done. Here is the diagram of the data flow among our components:

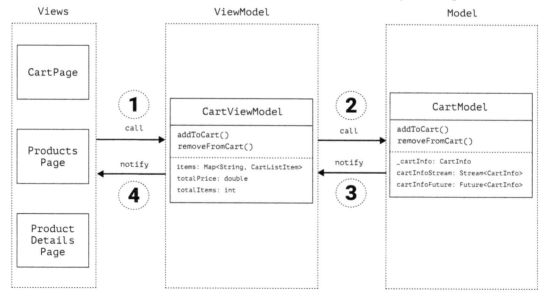

Figure 4.5 – Data flow among the Candy Store app components

Here, we have introduced `CartModel`, as well as renamed `CartNotifier` to `CartViewModel`. Here is how the data flows between them:

1. Our *Views* (`CartPage`, `ProductsPage`, and `ProductDetailsPage`) call the methods of `CartViewModel` on user interaction, such as when an item is added or removed from the cart. They do so by obtaining an instance of `CartViewModel` via `CartViewModelProvider` – an `InheritedWidget` widget that was introduced at the root of the tree.

2. Next, `CartViewModel` delegates this logic to `CartModel` by calling its methods.

3. Then, `CartViewModel` subscribes to the changes of a field of the `CartInfo` type, which contains relevant data about the cart, via a `Stream` class instance called `cartInfoStream`. Whenever the `cartInfo` field of `CartModel` changes, `CartModel` flushes those changes to `cartInfoStream`, which notifies its listeners of these changes – in our case, `CartViewModel`.

4. When `CartViewModel` receives an update notification from the stream, it updates its local fields (for example, `items`, `totalPrice`, and `totalItems`) and notifies its listeners, such as `CartPage`, via the `notifyListeners()` method.

You might be thinking that we introduced more boilerplate and haven't added any real benefit. In reality, the *Model* can be anything: it can be an API client that goes to the internet and fetches data, it can be a database that makes queries to the system, and so on. Furthermore, that *Model* can be reused by other *ViewModels* and in other places. It can even change completely: during the initial stage of development, when the API is not yet available, we can use an in-memory implementation. Or later, we may want to support an offline mode. If we encapsulate the communication of these business logic changes in the *ViewModel*, then in the best-case scenario, there will be zero changes in the *View* itself, which achieves the goal that we had in the first place: separation of business and UI logic.

But the scenario that we have worked on so far still feels a bit… artificial. In the real world, if you work with async data, your requests will likely never be fulfilled immediately, so you'll want to make sure that the user knows that their request is being processed by seeing a loader indicator. Alternatively, the user may see that something has gone wrong and expect to see a descriptive error. Let's make sure we're handling these scenarios correctly.

Encapsulating state in CartState

First of all, let's create a new class – `CartState`. It will be the source of truth for the current state of the cart and it will be read by our *View* and updated by our *ViewModel*. Let's take a look at it:

lib/cart_state.dart

```
class CartState {
  final Map<String, CartListItem> items;
  final double totalPrice;
  final int totalItems;
```

```
  final bool isProcessing;
  final Exception? error;

  CartState({
    required this.items,
    required this.totalPrice,
    required this.totalItems,
    this.isProcessing = false,
    this.error,
  });

  CartState copyWith({
    Map<String, CartListItem>? items,
    double? totalPrice,
    int? totalItems,
    bool? isProcessing,
    Exception? error,
  }) {
    return CartState(
      items: items ?? this.items,
      totalPrice: totalPrice ?? this.totalPrice,
      totalItems: totalItems ?? this.totalItems,
      isProcessing: isProcessing ?? this.isProcessing,
      error: error,
    );
  }
}
```

Here, we have done several things:

1. First, we encapsulated all of the fields that we used before – that is, `items`, `totalPrice`, and `totalItems`.

2. Then, we added two new fields that represent the current status of our actions: `isProcessing` and a nullable `error` field.

3. Finally, we introduced a `copyWith` method that will allow us to create copies of the existing data object with updated fields, without modifying the underlying object.

To make this easier to understand, let's update our diagram so that it represents our changes:

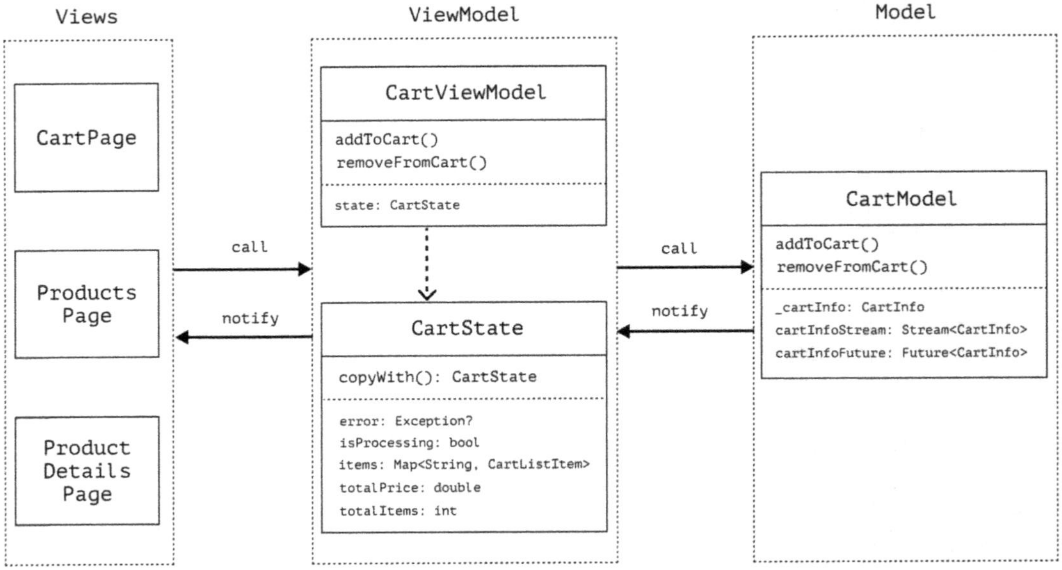

Figure 4.6 – Data flow among the Candy Store app components with CartState

Now, let's add the actual handling for the error and progress scenarios. First of all, we will need to modify our `CartModel` class in the following way:

lib/cart_model.dart

```
Future<void> addToCart(ProductListItem item) async {
  await Future.delayed(const Duration(seconds: 3));
  // throw Exception('Could not add item to the cart');
  // ... All code as before
}


Future<void> removeFromCart(CartListItem item) async {
  await Future.delayed(const Duration(seconds: 3));
  // throw Exception('Could not remove item from cart');
  // ... All code as before
}
```

A couple of things are going on here:

1. We changed the `void` type to `Future<void>` to introduce a delay that imitates a request to the real API.

2. We added a 3-second delay so that we can see it in the UI when we handle it.

3. In this case, the code is commented, but to test error handling, we will uncomment the code that throws exceptions.

Now that `CartModel` has been updated, let's introduce these changes in `CartViewModel`:

lib/cart_view_model.dart

```
class CartViewModel extends ChangeNotifier {
  final CartModel _cartModel = CartModel();

  CartViewModel() {
    _cartModel.cartInfoStream.listen((cartInfo) {
      // TODO: Should actually copy the Map and not just the
              reference, which we will do at the end of this chapter
      _state = _state.copyWith(
        items: cartInfo.items,
        totalPrice: cartInfo.totalPrice,
        totalItems: cartInfo.totalItems,
      );
      notifyListeners();
    });
  }

  CartState _state = CartState(
    items: {},
    totalPrice: 0,
    totalItems: 0,
  );

  CartState get state => _state;

  Future<void> addToCart(ProductListItem item) async {
    try {
      _state = _state.copyWith(isProcessing: true);
      notifyListeners();
      await _cartModel.addToCart(item);
```

```
      _state = _state.copyWith(isProcessing: false);
    } on Exception catch (ex) {
      _state = _state.copyWith(error: ex);
    }
    notifyListeners();
  }

  Future<void> removeFromCart(CartListItem item) async {
    // Same pattern as with `addToCart`
  }

  void clearError() {
    _state = _state.copyWith(error: null);
    notifyListeners();
  }
}
```

Let's observe what's going on here:

1. Instead of updating separate fields, we now always update the state object.

2. In the addToCart method, we now set the state to isProcessing when we start some action, as well as catch the error and set it to state if it happens. Because of this, our method has also become async and the return type has changed from void to Future<void>.

3. We introduced a new method called clearError that resets the error field of the state. We will use it to consume the error once we have addressed it in the UI.

Finally, let's see the changes that we need to do in CartPage:

lib/cart_page.dart

```
class _CartPageState extends State<CartPage> {
  late final CartViewModel _cartViewModel; // #1

  @override
  void initState() {
    super.initState();
    _cartViewModel = CartViewModelProvider.read(context); // #2
    _cartViewModel.addListener(_onCartViewModelStateChanged); // #3
  }

  @override
  Widget build(BuildContext context) {
```

```
    return Scaffold(
      appBar: AppBar(
        title: const Text('Cart'),
      ),
      body: ListenableBuilder(
        listenable: _cartViewModel,
        builder: (context, _) { // #7
          if (_cartViewModel.state.isProcessing) {
            // return view with a progress
          } else {
            // return view without a progress
          }
        },
      ),
    );
  }

  @override
  void dispose() {
    _cartViewModel.dispose(); // #6
    super.dispose();
  }

  void _onCartViewModelStateChanged() {
    // #4
    if (_cartViewModel.state.error != null) {
      _cartViewModel.clearError(); // #5
      ScaffoldMessenger.of(context).showSnackBar(
        const SnackBar(
          content: Text('Failed to perform this action'),
        ),
      );
    }
  }
}
```

Let's go through the changes step by step:

1. On line 1, we extracted CartViewModel so that it's an instance field of _CartPageState. By doing this, we can access it outside of the build method.

2. On line 2, we read CartViewModel in initState. As you may recall, previously, we accessed CartViewModelProvider via the .of method, and now it's .read. The difference in the implementation is that in the of method, we

accessed `InheritedWidget` via the `depend` method by running `context.dependOnInheritedWidgetOfExactType<CartViewModelProvider>()`. Remember, besides giving us a reference to the widget in question, it also subscribes the calling widget to the changes in `InheritedWidget`. We can't do that in `initState` because it is called only once per life cycle. So, to just read `InheritedWidget` without subscribing to its changes, `BuildContext` has another method: `context.getInheritedWidgetOfExactType<CartViewModelProvider>()`. This is why we're using it in the `read` method of `CartViewModelProvider`.

This is important!

When accessing `InheritedWidget`, you have two options:

You can rebuild the calling widget any time there are updates to the underlying `InheritedWidget`. In this case, make sure you use the getter that calls `context.dependOnInheritedWidgetOfExactType<CartViewModelProvider>()`. This can only be called in the methods that are invoked multiple times per life cycle, such as `build` or `didChangeDependencies`.

You only access `InheritedWidget` once and don't receive updates. In this case, make sure you use the getter that calls `context.getInheritedWidgetOfExactType<CartViewModelProvider>()`. It is safe to call this method from `initState`.

---

3. On line 3, we added a listener to `CartViewModel` (remember, it's a `ChangeNotifier` class and we can also add listeners to observe changes in it).

4. On line 4, we currently handle only the error state. We did this by showing a snack bar with the error text.

5. Once we consumed the error, we cleared it up via the `clearError` method that we just created. We did this on line 5.

6. To make sure that we didn't leak any resources, we `dispose` of our *ViewModel* in the state's `dispose` method. Under the hood, this method removes any listeners that we have added to `ChangeNotifier`. This happened on line 6.

7. Finally, we used the `isProcessing` flag to decide what kind of UI we wanted to build. If we want, we can even show the full-screen error if our `error` field is not `null`. This is up to the designs, but we now have all of the business logic to do that.

With this, we have achieved several things:

- Our business logic is decoupled from the UI logic and we can introduce changes to it without modifying the UI layer, as well as reuse our *ViewModel* with other kinds of *Views*, and swap out *Model* implementations without touching the *View*.

- Our app behavior is starting to look like a real application since it handles states such as progress and errors.

- We did all of this with the vanilla Flutter SDK, without using any third-party libraries.

If you want, you can stop here – you already have a working pattern, and especially if you're just starting, to not confuse yourself even more, you can stick to this. However, we will take this further. With the current approach, there are still a couple of issues:

- We need to add extra listeners to our *ViewModel* if we want to react to state changes without rebuilding the widget tree. This spreads the code all over the widget and makes it harder to navigate and understand what's going on and where.

- We need to manually care about when to dispose of the notifier as failing to remember to do this can lead to memory leaks.

- We need to rebuild the whole widget tree every time `notifyListeners` is called, even if there aren't any changes, or changes that our widget cares about.

- We still need to create `InheritedWidget` to provide our *ViewModel* down the tree.

- If we add more model resources with more streams to our *ViewModel*, it can become hard to handle because we need to manage all of the subscriptions and unsubscriptions, as well as potential race conditions.

All of these issues can be fixed by using a library called `flutter_bloc`. The pattern itself will seem familiar to you since we're already following it, but it also introduces tools and environments that will make all of this easier for us.

Implementing the MVVM and MVI patterns with flutter_bloc

First of all, what exactly is a *bloc*? **BLoC** is an acronym that stands for **Business Logic Component**. The general idea is very similar to all of the patterns that we have discussed so far, but the tooling and the naming are a bit different. Here, we have the *View*, which is our widget, we have our *Model* (also known as the data source), and we have the intermediary, *ViewModel*, which in the case of this library can be one of two things: a **cubit** or a **bloc**.

What is a cubit?

We will learn about both components of this library but we will start with the simpler one – the cubit. Let's take a look at the following diagram, which showcases its behavior:

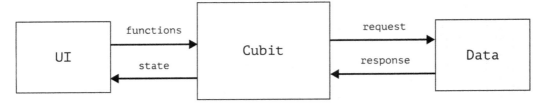

Figure 4.7 – Data flow among the Cubit components

Looks eerily familiar, right? That's because the idea, specifically regarding our implementation up to this point, is pretty much the same. The difference is in the mechanics, so let's take a look at how we can refactor the business logic code to use the cubit from the `flutter_bloc` library and how it can solve our existing problems.

First, let's look at an updated version of our components diagram:

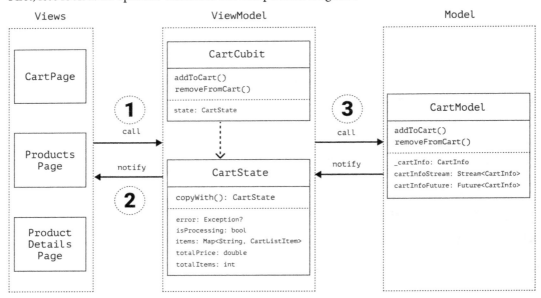

Figure 4.8 – Data flow among the Candy Store app components with CartCubit

The most notable thing here is that from this point of view, not much changes: we only rename `CartViewModel` to `CartCubit`. This is where we start seeing the difference between the pattern and the tooling: we are still following the same design pattern and modeling the same data flow.

What changes are the internal implementation details, something that's done with the help of the `flutter_bloc` library. Understanding patterns helps us select appropriate tools. We will dive into the details of each step shortly, but as an overview, here's what we will change:

1. Previously, we created an instance of a provider of `CartViewModel` – a custom `InheritedWidget` widget called `CartViewModelProvider`. The `flutter_bloc` library offers this functionality out of the box.

2. To notify our views of the changes in `CartState`, we used the API of `ChangeNotifier`, which included the `notifyListeners` method, `ListenableBuilder`, and custom listeners. Fortunately, `flutter_bloc` also offers a special API that enhances our experience.

3. In *Chapter 7*, we will learn how `flutter_bloc` assists in establishing a relationship between our *ViewModel* and *Model* layers.

Now, let's introduce `CartCubit`:

lib/cart_cubit.dart

```
class CartCubit extends Cubit<CartState> { // #1
  final CartModel _cartModel = CartModel();

  CartCubit()
      : super(
          const CartState(
            items: {},
            totalPrice: 0,
            totalItems: 0,
          ), // #2
        ) {
    _cartModel.cartInfoStream.listen(
      (cartInfo) {
        emit( // #3
          state.copyWith(
            items: cartInfo.items,
            totalPrice: cartInfo.totalPrice,
            totalItems: cartInfo.totalItems,
          ),
        );
      },
    );
  }

  Future<void> loadCart() async {
    try {
```

```
        emit(state.copyWith(isProcessing: true)); // #4
        final cartInfo = await _cartModel.cartInfoFuture;
        // TODO: Should actually copy the Map and not just the
                    reference, which we will do at the end of this chapter
        emit(
          state.copyWith(
            items: cartInfo.items,
            totalPrice: cartInfo.totalPrice,
            totalItems: cartInfo.totalItems,
          ),
        );
        emit(state.copyWith(isProcessing: false));
      } on Exception catch (ex) {
        emit(state.copyWith(error: ex));
      }
    }

  Future<void> addToCart(ProductListItem item) async {
      try {
        emit(state.copyWith(isProcessing: true));
        await _cartModel.addToCart(item);
        emit(state.copyWith(isProcessing: false));
      } on Exception catch (ex) {
        emit(state.copyWith(error: ex));
      }
    }

  Future<void> removeFromCart(CartListItem item) async {
      try {
        emit(state.copyWith(isProcessing: true));
        await _cartModel.removeFromCart(item);
        emit(state.copyWith(isProcessing: false));
      } on Exception catch (ex) {
        emit(state.copyWith(error: ex));
      }
    }

  void clearError() {
      emit(state.copyWith(error: null));
    }

  @override
  Future<void> close() async {
```

```
    _cartModel.dispose(); // #5
    super.close();
  }
}
```

Not much has changed in terms of structure, but there are some differences:

- First of all, instead of extending `ChangeNotifier`, we now extend `Cubit`. Pay attention to how we also specify `CartState` in the angle bracket that we will be working with. Here, `Cubit` supports the state out of the box.

- It also expects the state from the moment of its creation so that it always has relevant information and the consumer of this state can rely on it. To do that, we need to pass the initial state to the `super` constructor. We can do this internally, or if we want, we can allow the callers of our constructor to pass their own value. This is up to us and our use case.

- Here, `Cubit` has a special function that we need to call to notify our subscribers of state changes. Similarly, we called `notifyListeners` on `ChangeNotifier`, but now, we need to call `emit` and also pass the state to this emitter. It will soon come in handy when we need to distinguish whether there were any actual state changes or not.

- Just to emphasize, every time we make changes to our state, we just call `emit` and pass the new state as the parameter. We don't need to call `notifyListeners` after any internal state changes as the `emit` function does everything.

- Finally, `Cubit` has its own `close` function where we can dispose of any resources that we have acquired.

To see what's changed, let's go to the widget side of things and refactor it:

lib/cart_page.dart

```
class _CartPageState extends State<CartPage> {
  late final CartCubit _cartCubit; // #1

  @override
  void initState() {
    super.initState();
    _cartCubit = context.read<CartCubit>(); // #2
    _cartCubit.loadCart();
  }

  @override
  Widget build(BuildContext context) {
    return Scaffold(
      appBar: AppBar(
```

```
        title: const Text('Cart'),
      ),
      body: BlocConsumer<CartCubit, CartState>( // #3
        listener: (context, state) { // #4
          if (state.error != null) {
            _cartCubit.clearError();
            ScaffoldMessenger.of(context).showSnackBar(
              const SnackBar(
                content: Text('Failed to perform this action'),
              ),
            );
          }
        },
        builder: (context, state) { // #5
          // return view based on the `state` param
        },
      ),
    );
  }
}
```

Let's review it step by step:

1. First, we have swapped our *ViewModel* for `Cubit`.

2. Next, we get the reference to this `Cubit` by calling a `read` method on the context. This method comes from the `flutter_bloc` library and removes the need to create our own `InheritedWidget` widget that provides cubits to the widget tree. We will see this in more detail right after this snippet.

3. The next big thing is the widget that we use to observe our `Cubit` state: the `flutter_bloc` library offers several of them for different use cases, but the most robust of them is `BlocConsumer`. First of all, we need to specify the type of cubit and the type of state we're working on in the constructor so that Dart can infer those types in the callbacks.

4. Next, there are two interesting parameters. The first is `listener`, which has `context` and `state` as parameters. It is invoked when any change to the state occurs, so we don't have to attach custom listeners as we did before. This is pretty convenient and all in one place. The second is `builder`, which has the same parameters but expects a widget to be returned from it – which can build based on the state. What's more, in cases when we don't want to rebuild the whole widget tree on *any* change, we can override the `listenWhen` and `buildWhen` parameters and control them from there, whether we want `listener` or `builder` to be invoked! For example, you don't want to rebuild when an error event happens, but you do want to react to it in the listener.

Last, but not least, how do we provide the instance of this `Cubit` to our widget tree?

There is a special tool for that too, called `BlocProvider`:

lib/cart_page.dart

```
BlocProvider<CartCubit>(
    create: (context) => CartCubit(),
    child: const CartPage(),
  );
```

In `BlocProvider`, we describe how to create the `Cubit` instance that we want, as well as the widget child, which in our case is `CartPage`. There's no need to create any extra classes such as `CartViewModelProvider`, which provides our *ViewModel* to the widget tree. What's more, `BlocProvider` is smart enough to close the instance of `cubit` (or as we will soon see, `bloc`) when the widget that has created it is removed from the widget tree. So, with this approach, we have solved four out of the five problems presented by the previous solution! We still have potential problems with the streams, and if the cubit becomes too big, meaning it will have a lot of methods modifying the state, it might become a little bit hard to handle, as well as be a bit too explicit on the implementation details of the calling widget. To address this, we have the parent of the cubit – the bloc.

Understanding the MVI pattern

Before we move on to the practical differences, we need to discuss what stands behind the Bloc. Up to this point, we have worked with the MVVM pattern, but recently, another pattern has been derived from it – the MVI, or **Model-View-Intent**. Let's review its differences with MVVM:

Without ViewModel:

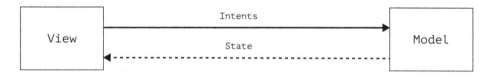

With ViewModel:

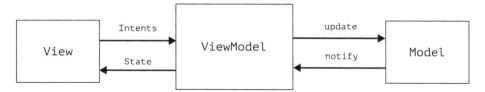

Figure 4.9 – Data flow among the MVI components

Once again, it is important to note that the pattern itself and specifically its interpretations in various frameworks and programming languages can vary. Here, we will discuss how it is applicable in the context of Flutter, especially with the help of the `flutter_bloc` library.

The main difference in MVI is the *Intent*. The concept of *Intent* introduces another layer of abstraction – instead of calling functions on our *ViewModel* itself, which we described as commands, we supply it with *Intents*, and the *ViewModel* resolves them itself, deciding how to react to them and what state to produce. While there is generally no *ViewModel* in MVI, it is still one of the variations of how to implement it, so we will stick to it.

MVI itself can be seen as an implementation of an even broader kind of architecture – event-based architecture. The idea is that some components produce events (such as UI triggers such as clicks, scrolls, taps, and so on) and some components consume and react to these events, such as *ViewModels*. The *ViewModel* (or any other dedicated component) transforms the events into a state. In MVI, these events are called *Intents*. And while we could add functionality to our cubit to implement this pattern, we don't need to – another component already does it for us, as well as handles all related problems. It's called the bloc.

Implementing MVI with BLoC

As opposed to the cubit, you shouldn't call methods on blocs. Instead, you should supply events via the `add` method of the bloc, and then implement their resolution in the bloc itself. First, let's visualize this:

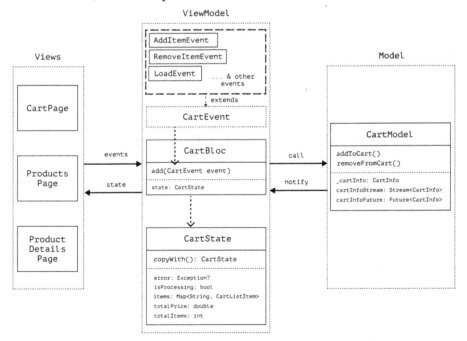

Figure 4.10 – Data flow among the Candy Store app components with CartBloc

At this stage, our diagram does finally change a little. Besides renaming CartCubit to CartBloc, we have also introduced a new class called CartEvent and its implementers: AddItemEvent, RemoveItemEvent, and so on. To map the MVI terminology to bloc terminology, the intents are represented by events. From the implementation's point of view, it means that now, instead of explicitly calling functions on CartBloc (as we did with CartCubit, such as addToCart and removeFromCart), we supply an instance of CartEvent via a single method – the add method of CartBloc – allowing it to resolve these events by itself. Now, it's time to implement it.

Let's take a look at what we need to change to implement this. First, we need to introduce the event class:

lib/cart_event.dart

```
sealed class CartEvent {
  const CartEvent();
}

final class Load extends CartEvent {
  const Load();
}

final class AddItem extends CartEvent {
  final ProductListItem item;

  const AddItem(this.item);
}

final class RemoveItem extends CartEvent {
  final CartListItem item;

  const RemoveItem(this.item);
}

final class ClearError extends CartEvent {
  const ClearError();
}
```

We will have the parent as a sealed class, which will act as the umbrella type for all of the events of this bloc, and a specific event per specific action, such as load, add, or remove. Before we move on to refactoring our cubit into a bloc, we need to make some updates to CartModel. Although we're applying these changes now, ideally they should have been done from the start, when introducing the CartModel, as technically they are not specific to MVI or BLoC, yet at that point it could have been overwhelming.

lib/cart_model.dart

```dart
class CartModel {
  CartModel._internal(); // #1

  static final CartModel _instance = CartModel._internal();

  factory CartModel() => _instance;

  final CartInfo _cartInfo = CartInfo(
    items: {},
    totalPrice: 0,
    totalItems: 0,
  );

  final StreamController<CartInfo> _cartInfoController =
      StreamController<CartInfo>.broadcast(); // #2

  Stream<CartInfo> get cartInfoStream => _cartInfoController.stream;

}
```

Here, we have done the following:

1. First, we created a singleton constructor so that it is possible to create only a single instance of `CartModel`. We did this so that the information in this model is consistent across the whole application. We will discuss creational patterns in much more detail in *Chapter 7*.

2. Then, we updated `StreamController` to `broadcast`. This is a special functionality of a stream in Dart, which allows the stream to be listened to by many listeners, instead of one. This is required if we create several instances of the same cubit.

> **Tip**
>
> The Streams API in Dart is very useful, and it's recommended to get familiar with it. To expand on the Streams API's capabilities, take a look at the `rxdart` library (`https://pub.dev/packages/rxdart`) which builds on top of it and makes the experience even more flexible. If you're familiar with RX programming from other languages, such as RxJava, you will find a lot of common APIs in RxDart too. However, pay attention to the differences related to multithreading. We will explore the multithreading aspect of Dart in more detail in *Chapter 9*.

Now, we need to update our cubit so that it's a bloc:

lib/cart_bloc.dart

```dart
class CartBloc extends Bloc<CartEvent, CartState> { // #1
  final CartModel _cartModel = CartModel();

  CartBloc()
      : super(
          const CartState(
            items: {},
            totalPrice: 0,
            totalItems: 0,
          ),
        ) {
    on<Load>(_onLoad); // #2
    on<AddItem>(_onAddItem);
    on<RemoveItem>(_onRemoveItem);
    on<ClearError>(_onClearError);
  }

  Future<void> _onLoad(Load event, Emitter emit) async {
    try {
      emit(state.copyWith(isProcessing: true));
      final cartInfo = await _cartModel.cartInfoFuture; // #3
      // TODO: Should actually copy the Map and not just the
      // reference, which we will do at the end of this chapter
      emit(
        state.copyWith(
          items: cartInfo.items,
          totalPrice: cartInfo.totalPrice,
          totalItems: cartInfo.totalItems,
        ),
      );
      emit(state.copyWith(isProcessing: false));
      await emit.onEach( // #4
        _cartModel.cartInfoStream,
        onData: (CartInfo cartInfo) {
          emit(
            state.copyWith(
              items: cartInfo.items,
              totalPrice: cartInfo.totalPrice,
              totalItems: cartInfo.totalItems,
```

```
          ),
        );
      },
      onError: (Object error, StackTrace stackTrace) {
        emit(
        state.copyWith(error: Exception('Failed to load cart
        info')));
      },
    );
  } on Exception catch (ex) {
    emit(state.copyWith(error: ex));
  }
}


Future<void> _onAddItem(AddItem event, Emitter emit) async {
  // Same as before
}


Future<void> _onRemoveItem(RemoveItem event, Emitter emit) async {
  // Same as before
}


void _onClearError(ClearError event, Emitter emit) {
  emit(state.copyWith(error: null));
}
}
```

Let's take a look at what is going on here:

1. Now, instead of extending `Cubit`, we extend `Bloc`. Note that in the diamond brackets, we also specify the parent class of the events that this bloc will handle.

2. In our constructor, we need to provide the `on` handlers for every type of event. This is very important because, as opposed to the cubit, we won't be calling methods on the bloc. The `on` handlers are implemented by the library and they let us control how to handle the events. Another feature that comes with these handlers are transformers – while we won't be diving into their details, keep in mind that with transformers, you can control all kinds of things: how events are handled (sequentially or asynchronously), whether to debounce the events or not, and so on.

3. When we load our bloc, we will read the initial state of our cart from the *Model* that we created previously.

4. After that, we subscribe to changes from the stream by using a special function of the bloc's `Emitter` – that is, `emit.onEach`. It handles all of the stream subscriptions and unsubscriptions for us and lets us deal with the data that we care about.

Finally, in our widgets, we need to change method calling to just calling add on the bloc and passing an event as the parameter. Here's an example in _CartPageState:

lib/cart_page.dart

```
@override
  void initState() {
    super.initState();
    _cartBloc = context.read<CartBloc>();
    _cartBloc.add(const Load());
  }
```

Instead of calling _cartCubit.load(), as we did previously, we're now using the add method of the bloc.

Note that this chapter is not intended to teach all of the details of how this (or any other) library works. The goal is to demonstrate patterns that are implemented by this library, how they can solve problems that we encounter, and where they stem from so that you can understand the hows and whys of the libraries you're using. Working code can be found in this book's GitHub repository; specifics of any libraries should be looked up in the relevant documentation at any given time.

Embracing the UDF pattern

So far, we have implemented various patterns, both by using what's available from the vanilla Flutter SDK, as well as the benefits offered by the flutter_bloc library. What we haven't talked about, but what we were still following in all of these patterns, is **unidirectional data flow** (UDF). The idea behind this pattern is that data has only one direction through which it flows. Following this pattern makes it easier to reason about state changes, control where these state changes are introduced, predict when those changes will happen, and avoid unwanted side effects. Both of our MVVM and MVI implementations with ChangeNotifier, Cubit, and Bloc embrace it – we never modify the state outside of the *ViewModel*; it is all encapsulated either in the *ViewModel* methods itself (in the case of ChangeNotifier and Cubit) or in the event handlers in case of Bloc. The state is also always emitted only from one single source of truth. Regardless of the details or specifics of the pattern that you select, making sure that it follows UDF will ensure that your state management is predictable and reliable. The best thing is that UDF plays very nicely with the declarative UI-building approach of Flutter since our UI is always the function of the state.

Implementing the Segmented State Pattern

So far, we have mostly been working with various implementations of the component that connects the *Model* to the *View*. We briefly looked at the state and even made it more life-like by handling progress and error via the isProcessing and error fields. Having distinct states for progress, error, and value is a very common pattern and it has even been given a name – the Segmented State Pattern or the Triple Pattern (`https://docs.flutter.dev/data-and-backend/state-mgmt/options#triple-pattern-segmented-state-pattern`). While there is a library with this name, I think that this name is very suitable for the pattern itself, regardless of how it is implemented.

While our implementation works for our example and does the job, in reality, it introduces quite some architectural problems. Let's talk about them.

First of all, it is a good idea and practice to make the existence of an invalid state that is not explicitly handled impossible. What is an invalid state? Well, for example, if we have `isProcessing` set to `true`, and the `error` field is not `null`, it leads to inconsistency – is it in progress? Is it an error? Is it an old error or a new error? Where is the truth? Another problem with this approach is that if we add any more fields to the state, we will also need to add their progress and error fields, which will become cumbersome and hard to manage:

lib/cart_state.dart

```
class CartState {
    final bool loadInProgress;
    final Exception? loadError;
    final CartInfo? cartInfo;
    final bool checkoutInProgress;
    final Exception? checkoutError;
    final T? checkoutValue;
}
```

One way we could fix this is by making the status an algebraic type, such as via an enum value:

```
enum CartStatus { initial, progress, error, success }

class CartState {
    final CartStatus loadStatus;
    final Exception? loadError;
    final CartInfo? cartInfo;
    final CartStatus checkoutStatus;
    final Exception? checkoutError;
    final T? checkoutValue;
}
```

While we can trust the status itself by using this approach, no one is stopping us from still making the state inconsistent by having a non-null error state. Plus, it is as cumbersome as before.

Our next attempt could be to use inheritance:

```
sealed class CartStatus {}

class CartIdleStatus extends CartStatus {}
class CartLoadStatus extends CartStatus {}
class CartErrorStatus extends CartStatus {
    final Exception? error;
}
class CartSuccessStatus extends CartStatus {
    final CartInfo? cartInfo;
}

class CartState {
    final CartStatus cartStatus;
}
```

With this approach, we solve the inconsistency problem because there is always only one single cartStatus, but we also introduce a lot of boilerplate. Imagine creating five classes for each possible data state. This would bloat the code base unnecessarily very fast, and it would still be hard to manage. Since the pattern is generally the same and only the actual data changes, we can extract it into a reusable class. We will call it DelayedResult:

lib/delayed_result.dart

```
class DelayedResult<T> extends Equatable {
  final T? value;
  final Exception? error;
  final bool isInProgress;

  const DelayedResult.fromError(Exception e)
      : value = null,
        error = e,
        isInProgress = false;

  const DelayedResult.fromValue(T result)
      : value = result,
        error = null,
        isInProgress = false;

  const DelayedResult.inProgress()
      : value = null,
        error = null,
        isInProgress = true;
```

```
const DelayedResult.idle()
    : value = null,
      error = null,
      isInProgress = false;

bool get isSuccessful => value != null;

bool get isError => error != null;

bool get isIdle => value == null && error == null && !isInProgress;

@override
List<Object?> get props => [value, error, isInProgress];
}
```

We create such a class once and then we can reuse it every time in all of our notifiers, cubits, or blocs. You can use the class as-is in your apps or follow its evolution from the source (`https://github.com/ChiliLabs/dart-delayed-result`). For example, if we refactor `CartBloc`, it could look like this:

lib/cart_state.dart

```
class CartState {
  final Map<String, CartListItem> items;
  final double totalPrice;
  final int totalItems;
  final DelayedResult<void> loadingResult;
}
```

Now, instead of the `isProcessing` and `error` fields, we have one – the `loadingResult` field. We could even go as far as to remove the other fields and make `DelayedResult` not `void` but `CartInfo`. However, we won't do that because we want to be able to show an error in our UI while also showing the cached results. Now, in our event handling logic, we can do something like this:

lib/cart_bloc.dart

```
Future<void> _onAddItem(AddItem event, Emitter emit) async {
    try {
      emit(
        state.copyWith(
          loadingResult: const DelayedResult.inProgress(),
        ),
      );
```

```
      await _cartModel.addToCart(event.item);
      emit(state.copyWith(loadingResult: const DelayedResult.idle()));
    } on Exception catch (ex) {
      emit(
        state.copyWith(
          loadingResult: DelayedResult.fromError(ex),
        ),
      );
    }
  }
```

Instead of setting different fields for progress and errors, we now manipulate the status with just one, and it's always consistent with the actual state of things.

Avoiding bugs related to state equality and mutability

One last thing that we need to discuss in this chapter is the equality and immutability of data classes, specifically the state. What you need to know about the equality of objects in Dart is that, by default, they're only equal if they're *actually* the same instance. Let's compare objects in vanilla Dart:

```
void main() {
  final o1 = Object();
  final o2 = Object();
  final o3 = o1;
  print(o1 == o2); // prints "false"
  print(o3 == o1); // prints "true"
}
```

Even though o1 and o2 look the same visually, in computer memory, they are two different objects, hence why the == operator returns false by default. We could override that behavior by manually overriding the == operator in every data class, but that would be a lot of boilerplate. Instead, many libraries do this for us. The one that we will be using is called **Equatable** (https://pub.dev/packages/equatable). The idea is that your data class extends Equatable and overrides the props list. In this list, you specify which fields you want to compare your data class by, and under the hood, Equatable overrides the == operator based on these props. This is crucial for the builder and listener methods of BlocConsumer to work efficiently and avoid unnecessary rebuilds. Those methods check the result of == operator, and in case the objects are different, emit a new state. Dart objects are always different by default, hence those comparisons won't work without overriding the == operator. Let's take a look at what CartState looks like with Equatable:

lib/cart_state.dart

```
class CartState extends Equatable {
```

```dart
  final Map<String, CartListItem> items;
  final double totalPrice;
  final int totalItems;
  final DelayedResult<void> loadingResult;

  const CartState({
    required this.items,
    required this.totalPrice,
    required this.totalItems,
    required this.loadingResult,
  });

  CartState copyWith({...}); // as before

  @override
  List<Object?> get props => [
      items,
      totalPrice,
      totalItems,
      loadingResult,
    ];
}
```

In our class definition, we are extending Equatable and in the overridden props field, we specify all of the fields we care about.

One last thing we need to mention is data model immutability. It is good practice to make data models immutable. This means that once you create an instance of a data object, you can't change any of its fields. Technically, this can be implemented by marking all of the fields with the final keyword. This approach offers several benefits. For one, your data model is always consistent and predictable because to change it, you need to explicitly make a new instance of it. For that, we have convenience methods such as copyWith, which creates a new instance of the class but only allows you to change the fields that you want, copying the other fields from the original object.

An important thing to note here is how Dart passes arguments to functions. Technically, Dart is a pass-by-value language. This means that any changes that are made to the argument inside the function won't affect the original variable. However, in Dart, all data types are essentially references. This means that instead of copying the entire object, the function receives a copy of the reference to the object in memory. In practice, this means two things:

1. While the function receives a copy of the reference, it still points to the same object in memory. So, modifications to the object's properties will be reflected in the original variable.

2. If the variable holding the reference is reassigned, it will stop pointing to the same object in memory, so further changes won't be applied to the original object.

Therefore, if you modify that object inside the function, the changes will be visible outside the function as well. This can be a big problem with collections, especially in the state. Let's take a look at an example:

```
void main() {
  List<int> numbers = [1, 2, 3];
  modifyList(numbers);
  print(numbers); // prints: [100, 2, 3]
  reassignList(numbers);
  print(numbers); // also prints: [100, 2, 3]
}
void modifyList(List<int> numbers) {
  numbers[0] = 100; // changes the original list
}
void reassignList(List<int> numbers) {
  numbers = [100, 200, 300]; // reassigns the local variable
}
```

Even though we passed the original numbers value to modifyList as a parameter, we have mutated the underlying list. This means that if you were accessing that list anywhere else in the code, it would also be modified there. So, whenever you're passing around collections, to ensure the preservation of immutability, make sure you make a copy of the underlying collection. Unfortunately, this problem hasn't been perfectly resolved in Dart yet. You could copy a collection in many ways, such as by calling List.of(numbers).

Before we wrap up this section, let's take a look at how object mutability and incorrect equality handling can lead to bugs in our Candy Store app. Our current implementation works fine, but it has a hidden bug. In CartModel, we modify _cartInfo.items instead of creating a copy and emit this same collection to CartBloc. If we make a couple of modifications to CartBloc, we will encounter the following bug:

lib/cart_bloc.dart

```
Future<void> _onAddItem(AddItem event, Emitter emit) async {
  try {
    // #1
    //emit(
    //  state.copyWith(
    //    loadingResult: const DelayedResult.inProgress(),
    //  ),
    //);
    await _cartModel.addToCart(event.item);

    // #2
```

```
        //emit(
        //   state.copyWith(
        //     loadingResult: const DelayedResult.idle(),
        //   ),
        //);
    } on Exception catch (ex) {
      emit(
        state.copyWith(
          loadingResult: DelayedResult.fromError(ex),
        ),
      );
    }
  }
}
```

Let's review what we have done:

1. On lines 1 and 2, we have commented out emitting state changes to `inProgress` and `idle`. This way, when we call `_onAddItem`, the state object will stay the same and the UI won't change.

2. On lines 3 and 4, we commented out emitting the changes to the `totalPrice` and `totalItems` fields, but we have kept changes to the `items` field.

3. As a result, our UI won't be rebuilt at all! From Dart's perspective, `cartInfo.items` point to the same object as before, which makes them equal, so Bloc doesn't see state changes and the widget isn't rebuilt.

4. If that wasn't enough, nothing is stopping us from adding or removing items from the underlying map by calling methods such as `clear` or `addAll`. This can lead to very obscure and hard-to-trace bugs.

So, how can we fix these issues? There are many ways to do this, some of them more correct than others. For example, in `CartModel`, we can keep the mutable version of `CartInfo` for convenience, but emit a copy of it to the listeners. When doing so, we can also make a copy of our cart items map with a special constructor called `Map.unmodifiable`. Not only will it make a copy of the map, which will solve our equality problem, but will also throw runtime errors if we try to modify the underlying collection. Let's take a look at the updated code in `CartModel`:

lib/cart_model.dart

```
final CartInfo _cartInfo = CartInfo(
    items: {},
    totalPrice: 0,
    totalItems: 0,
  );
```

```
// 1
Future<CartInfo> get cartInfoFuture async => _cartInfo.copyWith(
    items: Map.unmodifiable(_cartInfo.items),
  );

Future<void> addToCart(ProductListItem item) async {
  // Code omitted for demo purposes,
  // modify _cartInfo as before
  // 2
  final cartInfo = _cartInfo.copyWith(
    items: Map.unmodifiable(_cartInfo.items),
  );

  // 3
  _cartInfoController.add(cartInfo);
}

Future<void> removeFromCart(CartListItem item) async {
  // Code omitted for demo purposes,
  // modify _cartInfo as before
  final cartInfo = _cartInfo.copyWith(
    items: Map.unmodifiable(_cartInfo.items),
  );
  _cartInfoController.add(cartInfo);
}
```

In this piece of code, we did the following:

1. From cartInfoFuture, we returned a copy of _cartInfo by calling copyWith and explicitly using Map.unmodifable to create an immutable copy of the map.

2. We do the same thing whenever we modify _cartInfo in add or remove operations.

3. We emitted the copy via _cartInfoController instead of the original.

By doing this, we have fixed our bugs. It is crucial to be mindful when working with collections in Dart, especially when they're mutable. You can review the compilable step-by-step reproduction and fix of these issues in the CH04/final folder and its commit history (https://github.com/PacktPublishing/Flutter-Design-Patterns-and-Best-Practices/tree/master/CH04/final/candy_store).

Summary

In this chapter, we delved even deeper into the realm of state management. First, we explored the history of how popular state management patterns came to be. We learned the ideas behind such patterns as MVC, MVVM, MVI, and BLoC, as well as how to implement them in Flutter both by using tools available out of the box and via a well-known and widely used library called `flutter_bloc`. We also implemented a pattern called Segmented State with `DelayedResult`, which applies to any other solution that we choose. Moreover, we learned about important concepts in Flutter and Dart, such as Dart streams and object equality, as well as how to manage equality with the help of the `Equatable` library.

Now, we're equipped with the tools we need to manage the state of our Flutter applications, which vary in terms of complexity. We have the required knowledge to decide on the most suitable pattern and tools for ourselves since there is no single pattern that suits everyone.

In *Chapter 5, Creating Consistent Navigation*, we will explore the topic of navigation in Flutter. We will review the difference between declarative and imperative approaches to navigation, their pros and cons, as well as their application in our Flutter projects.

Get this book's PDF version and more

Scan the QR code (or go to `packtpub.com/unlock`). Search for this book by name, confirm the edition, and then follow the steps on the page.

Note: Keep your invoice handy. Purchases made directly from Packt don't require an invoice.

5

Creating Consistent Navigation

Navigating in a Flutter app is like finding your way through a new city. Flutter gives us a bunch of tools to move smoothly from one screen to another, making sure users always find what they're looking for. Whether your app is simple or complex, knowing how to guide users properly is key.

In this chapter, we're diving into how to use Flutter's navigation tools. We'll start with the basics, such as routes and navigators, and then explore more advanced stuff for bigger apps. You'll learn how to connect parts of your app in a way that feels natural and easy.

We're also going to look at deep linking, which lets users jump into your app from links outside it. This is super handy for sharing and getting around in your app. Plus, we'll talk about how to pass data around between screens. This means you can keep your app running smoothly and make sure users have a great experience.

By the end of this chapter, you'll have a solid understanding of navigation in Flutter. You'll get to try out what you've learned through practical exercises, giving you the confidence to build apps that are not just functional but also user-friendly.

We will cover the following topics in this chapter:

- Getting started with navigation in Flutter
- Leveling up with advanced navigation techniques
- Comparing Navigation 1.0 and 2.0

Technical requirements

To get started with this chapter, you'll need to check out the CH05 initial branch. From there, your journey will span multiple aspects of navigation in Flutter. Here is your initial branch to give it a start:

- Here are your initial branch/files to start to this chapter: `https://github.com/PacktPublishing/Flutter-Design-Patterns-and-Best-Practices/tree/master/CH05/initial/candy_store`.

- Here is the final version of imperative navigation implementations we will do: `https://github.com/PacktPublishing/Flutter-Design-Patterns-and-Best-Practices/tree/master/CH05/final_navigation_1/candy_store`.

- Here is the final version of Navigation 2.0 in a single file: `https://github.com/PacktPublishing/Flutter-Design-Patterns-and-Best-Practices/blob/master/CH05/final_navigation_2_declarative/candy_store/lib/main.dart`. This is an alternative navigation, intended for further learning. It is not implemented in the main app.

Getting started with navigation basics in Flutter

In this section, we'll learn the main terms of navigation and how to use Flutter's navigation tools effectively. Navigation is super important for providing a smooth user experience, guiding users through your app's features with ease. Let's dive in and start with the basics. There are two types of navigation in Flutter: **imperative** and **declarative** navigation. We will go through both and expand on them to learn more. First, let's learn the basics.

Understanding Navigator 1.0 and routes

Everything you're about to learn in this segment is referred to as navigation 1.0, or imperative navigation. This is where your journey in mastering app navigation starts.

Flutter utilizes a component known as the **Navigator** to keep track of the screens or pages within your app. You can think of the Navigator as a tour guide in a new city, guiding you from one location to another. These locations are represented by a series of **route** objects. To move to a different screen, the Navigator employs methods such as `push()` to add a route to the navigation stack and `pop()` to remove the current route, bringing you back to the previous screen.

Navigation can be managed using either **named routes** or **anonymous routes**. An example of this, which you might remember from *Chapter 3*, involves opening the `CartPage`. This action is facilitated by Navigation 1.0 (imperative style) through the use of anonymous routing. Let's revisit the code snippet found in `main_page.dart`:

```
void openCart() {
  Navigator.of(context).push(
    MaterialPageRoute(
      builder: (context) => CartPage.withBloc(),
    ),
  );
}
```

Both `MaterialApp` and `CupertinoApp` include a Navigator by default, and this piece of code demonstrates how to directly create and navigate to a new screen using `MaterialPageRoute`. It is an object that is a subclass of Route and gives it a basic screen transition animations covering material design principles.

Returning to a previous screen with the pop() Method

After navigating to `CartPage` using `push`, you might want to return to the previous screen. This is where the `pop()` method comes into play. Let's add a button to go back:

```
ElevatedButton(
    onPressed: () => Navigator.of(context).pop(),
    child: const Text('Go Back!'),
),
```

In this snippet, tapping the **Go Back** button triggers the `pop()` method. This action effectively removes the `CartPage` from the navigation stack, bringing the `MainPage` back into view.

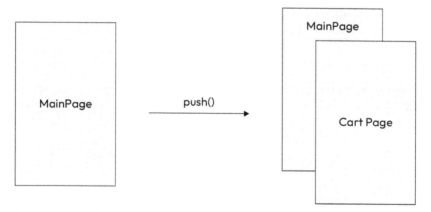

Figure 5.1 – A diagram showing how the push() command helps with navigation within the app

Figure 5.1 shows how the pages are stacked. When you use the `push()` method, the `CartPage` widget is placed on top of `MainPage`, and `MainPage` is still part of the widget tree. When you `pop()`, you can go back to `MainPage`.

Now that you have learned about anonymous navigation, let's check named routes.

Exploring named routes in Flutter navigation

Named routes offer us another approach for navigating between screens. This method allows you to assign a unique string identifier to each route or screen in your app, aiming to simplify the navigation process and make your code cleaner and more maintainable. However, currently, as can be found in the official documentation (`https://docs.flutter.dev/ui/navigation#using-named-routes`), it is not recommended for most applications. Before digging into the reasons, let's learn about named routes usage.

Defining routes in MaterialApp

First, we need to define our routes in our `MaterialApp`:

```
initialRoute: '/',
    routes: {
      '/': (context) => const MainPage(),
      '/cart': (context) => CartPage.withBloc(),
    },
```

In this configuration, `initialRoute` is set to `/`, directing the app to open at the `MainPage` when it starts. The routes map then defines two paths: the `/` home route and a `/cart` route for `CartPage`. This setup is crucial for employing named routes in your app.

Adapting the navigation to use named routes

Next, we adjust our navigation function to utilize these named routes. By updating the `openCart` function in `main_page.dart`, we can navigate to the `CartPage` using its route name:

```
void openCart() {
  Navigator.of(context).pushNamed('/cart');
}
```

This method leverages the route names defined in the `MaterialApp` to navigate between screens, showcasing a simple and effective way to implement named routing.

Understanding the limitations of named routes

While named routes offer a streamlined approach to navigation, there are important considerations and limitations to be aware of, as highlighted by the official Flutter documentation (`https://docs.flutter.dev/ui/navigation#using-named-routes`):

- **Lack of customization for deep links**: Named routes can manage deep links, but they do not allow for customized behavior. This means that the app might not consider the user's current navigation state when a deep link is opened, potentially leading to a disjointed user experience.

- **No support for browser forward button**: For web applications, named routes do not support using the browser's forward button. This limitation can impact the navigation experience in web-based Flutter apps, as users might expect these buttons to work as they do in traditional web applications.

Given these constraints, while named routes are efficient for straightforward navigation needs, they may not suit complex apps, particularly those requiring sophisticated deep linking or full utilization of web browser navigation features. The upcoming sections will explore advanced navigation options that address these challenges, offering more flexibility and control over the app's navigation behavior.

Now that we've got a good grasp of the basics, let's move on to some more advanced techniques.

Leveling up with advanced navigation techniques

After we learn about Navigator 2.0, which we also call declarative navigation, using the simpler, command-based navigation might feel limiting. It's hard to add or remove many pages at once or to deal with a page below the current one in the navigation stack. However, if you're content with the simpler version, it's still fine and works well. *The existence of Navigator 2.0 doesn't mean that the earlier version is outdated or unusable.*

As Flutter apps become more complex, you might need more advanced ways to move around the app. Advanced navigation, such as Navigator 2.0, allows you to change routes dynamically. Making your own route changes gives you strong tools to move through your app more easily. Let's look at these ideas and how they can make your app more enjoyable for users.

> Reminder
> Make sure that you're starting fresh from the initial version found in the GitHub repository. We'll build everything from scratch based on that version. The work done in the previous section is no longer needed. You will see in the app that we will use the previous section's code and keep building on top of it. Navigator 2.0 is built in the `main.dart` file in the repository and this example is for you to learn, discover, and be able to apply on the way.

Navigator 2.0 and declarative routing

Navigator 2.0, or the Router API, handles the app navigation by basing it on the app's overall state. This shift means the displayed screens and routes are determined by the app's current state, allowing for a more dynamic and responsive navigation experience. This approach is particularly beneficial for handling deep links and integrating with web URLs seamlessly. To fully grasp the capabilities of Navigator 2.0, let's familiarize ourselves with some key concepts:

- `Page`: Acts as an immutable object that forms the navigator's history stack, representing each screen within your app

- `Router`: Determines the active list of pages based on the app's state or platform changes, effectively deciding what content to show at any given moment

- `RouteInformationParser`: Transforms incoming route information into a user-defined data type, facilitating the interpretation of URLs or navigation actions

- `RouterDelegate`: Plays a crucial role in managing the app's navigation state, responding to changes detected by the `RouteInformationParser` and updating the navigator accordingly.

- `BackButtonDispatcher`: Manages the behavior of the back button within your app, ensuring that navigation actions are correctly handled across different platforms; tells the `Router` about back button presses.

Creating an example with Navigator 2.0

We're about to explore how everything connects in Navigator 2.0, focusing on a web app using Chrome. This way, you'll see how the URL parser works, how changes in the URL affect the app, and vice versa.

We're beginning anew. Previous chapters were introductions and adapting them to Navigator 2.0 would be too complex. So, we're turning to a new page to dive into Navigator 2.0. Yes, we'll continue with our dessert theme.

Here's the plan:

- Listing desserts and allowing users to select them for more details

- Enabling navigation back to the list using the browser's back button, mirroring traditional web experiences

- Dynamically updating the browser's URL to reflect the current page, enhancing shareability and user orientation within the app

- Handling invalid URLs by displaying an error page, ensuring robustness in navigation

Creating navigation2_main.dart:

To start with Navigator 2.0 in your Flutter project, follow these steps:

1. Open your Flutter project.
2. Navigate to the `lib` folder.
3. Create a new file and name it `navigation2_main.dart`.
4. Create an app called `Candy Shop`.

Here's a simple setup for the `Candy Shop` app using `MaterialApp`:

```
MaterialApp(
      title: 'Candy Shop',
      home: Navigator(
        pages: const [
          MaterialPage(
            key: ValueKey('DessertsPage'),
            child: Scaffold(
              body: Center(
                child: Text('Welcome to Candy Shop!'),
              ),
            ),
          )
        ],
        onPopPage: (route, result) => route.didPop(result),
      ),
    );
```

The preceding code is pretty much like any simple app that you've already created. As you can see, the `Navigator` has `pages` arguments. So, once we change the `Page` object, we should see the stack of routes that should adapt and match. Let's build the app to show the list of desserts first.

Defining the dessert class

We'll start by defining a `Dessert` class to model our data, then update the app to display a list of these desserts. First, let's define the `Dessert` class with the `name`, `description`, and `imageUrl` properties. This class will allow us to create and use dessert objects throughout our app:

```
class Dessert {
  final String name;
  final String description;
  final String imageUrl;
  Dessert(this.name, this.description, this.imageUrl);
}
```

With the `Dessert` class in place, we can now proceed to populate our app with dessert data and display a list of delicious options to the user. This approach keeps our app organized and makes it easy to work with dessert data throughout the application.

Managing dessert selections and list

Within the `_CandyShopAppState`, we'll maintain a list of available desserts and track any dessert that's currently selected by the user:

```
Dessert? _selectedDessert;
  List<Dessert> desserts = [
    Dessert(
      'Cupcake',
      'A delicious cupcake with a variety of flavors and toppings',
      'resources/images/cupcake.webp',
    ),
    Dessert(
      'Donut',
      'A soft and sweet donut, glazed or filled with your favorite
flavors',
      'resources/images/donut.webp',
    ),
    Dessert(
      'Eclair',
      'A long pastry filled with cream and topped with chocolate
icing',
      'resources/images/eclair.webp',
    ),
```

Next, we'll look at displaying lists.

Displaying the list of desserts

Now, we want to show the list of desserts in a screen, and once a user has tapped to add any dessert, it should update the `_selectedDessert` value. We can call that function if we integrate the `onTapped` functionality into our `Widget`.

Here is a basic `DessertsListScreen` to list the `Dessert`:

```
class DessertsListScreen extends Stateless Widget {
  final List<Dessert> desserts;
  final ValueChanged<Dessert> onTapped;

  const DessertsListScreen({
    super.key,
    required this.desserts,
    required this.onTapped,
  });
```

```
  @override
  Widget build(BuildContext context) {
    return Scaffold(
      appBar: AppBar(
        title: const Text('All Desserts'),
      ),
      body: ListView(
        children: [
          for (var dessert in desserts)
            ListTile(
              trailing: ClipRRect(
                borderRadius: BorderRadius.circular(8),
                child: SizedBox(
                  width: 72,
                  height: 72,
                  child: Image.asset(dessert.imageUrl),
                ),
              ),
              title: Text(dessert.name),
              subtitle: Text(dessert.description),
              onTap: () => onTapped(dessert),
            )
        ],
      ),
    );
  }
}
```

Each dessert in the list is represented by a ListTile, displaying the dessert's image, name, and description. When a user taps on a dessert, the onTapped function is invoked, updating the _selectedDessert with the chosen dessert. This interaction is key for moving to a detailed view of the selected dessert.

Integrating the DessertsListScreen into our app

Now we will replace our almost empty Scaffold with the DessertsListScreen we just created:

```
MaterialPage(
          key: const ValueKey('DessertsPage'),
          child: DessertsListScreen(
            desserts: desserts,
            onTapped: _handleDessertTapped,
          ),
        )
```

Then, we can proceed with the onTapped functionality that we discussed previously.

The _handleDessertTapped function is invoked upon the selection of a dessert. It updates the _selectedDessert state to mirror the user's choice:

```
void _handleDessertTapped(Dessert dessert) {
  setState(() {
    _selectedDessert = dessert;
  });
}
```

After selecting a dessert, the app's state changes by assigning the chosen dessert to _selectedDessert. However, this change has not yet resulted in any visual update in the app, as the code to display the dessert details has not been implemented.

Creating the DessertDetailsScreen for detailed views

To visually represent the details of the selected dessert, we should create a DessertDetailsScreen. This screen aims to provide a detailed view, showcasing the dessert's image, name, and description:

```
Scaffold(
  appBar: AppBar(
      title: Text('Detail Page for ${dessert.name}'),
    ),
      body: Center(
        child: Column(
          crossAxisAlignment: CrossAxisAlignment.center,
          children: [
            [
              ClipRRect(
                borderRadius: BorderRadius.circular(8),
                child: SizedBox(
                  width: 250,
                  height: 250,
                  child: Image.asset(dessert.imageUrl),
                ),
              ),
              Text(dessert.name, style: Theme.of(context).textTheme.
titleLarge),
              Text(dessert.description, style: Theme.of(context).
textTheme.titleMedium),
            ],
          ],
```

```
            ),
        );
    }
}
```

This is also another simple page that has only an `AppBar`, and shows the dessert's picture, name, and description. Now, we need to adapt our navigation to be able to show this page once the `_selectedDessert` is updated:

```
if (_selectedDessert != null)
    MaterialPage(
        key: ValueKey(_selectedDessert),
        child: DessertDetailsScreen(
        dessert: _selectedDessert!,
         ),
        )
```

This code snippet, positioned right after our initial `MaterialPage` setup, checks for a selected dessert. If a dessert is chosen, the app proceeds to render the `DessertDetailsScreen`, passing the selected dessert as an argument. This setup ensures a smooth navigation experience from the `DessertsListScreen` to the `DessertDetailsScreen` and back, facilitated by the unique `ValueKey` associated with each selected dessert, thus enhancing the app's navigation reliability.

Synchronizing app state with browser's URL

While our navigation setup effectively manages screen transitions, you may notice that the app's state changes aren't reflected in the browser's URL. To align the app's navigation state with the browser's URL, introducing a Router becomes essential.

To integrate the Router, we need to develop the following components:

- `RouteInformationParser`: Parses the incoming route information into a user-defined type, aiding in the interpretation of browser URLs

- `RouterDelegate`: Manages the app's navigation state in response to route information changes and is pretty important for updating the displayed content

These components collaborate to parse route information and adjust the app's state, ensuring the navigation remains in sync with the browser's URL.

Structuring navigation with router and route paths

To make our Flutter application more versatile and web-friendly, we incorporate structured navigation states and a Router.

The `DessertRoutePath` class plays a huge role in representing the different navigation states within our app:

- **Home Route**: Marks the app's landing page, displaying a list of desserts
- **Details Route**: Occurs when a user selects a dessert to view more about it, identified by a unique ID
- **Unknown Route**: Engaged when a navigation attempt does not match any predefined paths, leading to an error page

Check out the code:

```
class DessertRoutePath {
  final int? id;
  final bool isUnknown;

  DessertRoutePath.home()
      : id = null,
        isUnknown = false;

  DessertRoutePath.details(this.id) : isUnknown = false;

  DessertRoutePath.unknown()
      : id = null,
        isUnknown = true;

  bool get isHomePage => id == null && !isUnknown;
  bool get isDetailsPage => id != null;
}
```

We will handle all routes in the app with a single class. For advanced apps, you can use different classes to implement a superclass or manage route information in your own way. This setup not only simplifies the management of navigation states but also aligns the app's internal navigation with web URL standards, supporting direct navigation to pages via URLs. Now we can go for implementing the `RouterDelegate`.

This delegate is responsible for the following:

- Rendering the current route as a widget

- Reacting to changes in the route path and updating the app's state and UI accordingly

- Managing the navigator key, which is essential for identifying the `Navigator` widget this delegate is working with

Here's a simplified version of what these components look like in code, with placeholders as `UnimplementedError` for future implementation:

```
class ExampleRouterDelegate extends RouterDelegate<ExampleRoutePath>
    with ChangeNotifier,
PopNavigatorRouterDelegateMixin<ExampleRoutePath> {

  @override
  Widget build(BuildContext context) {
    throw UnimplementedError();
  }

  @override
  GlobalKey<NavigatorState> get navigatorKey => throw
UnimplementedError();

  @override
  Future<void> setNewRoutePath(ExampleRoutePath configuration) {
    throw UnimplementedError();
  }
```

The preceding code snippet is an `ExampleRouterDelegate` for you to see purely the base of it.

Create the DessertRouterDelegate for dynamic navigation

However, in our code, we can use `DessertRoutePath`'s list of desserts, whether we want to show a 404 page or not. All these data and details can be kept under `RouterDelegate`:

```
class DessertRouterDelegate extends RouterDelegate<DessertRoutePath>
    with ChangeNotifier,
PopNavigatorRouterDelegateMixin<DessertRoutePath> {

  @override
```

```
    final GlobalKey<NavigatorState> navigatorKey;

    Dessert? _selectedDessert;
    List<Dessert> desserts = ourListOfDessertsBefore;

    DessertRouterDelegate() : navigatorKey =
GlobalKey<NavigatorState>();
}
```

Let's check the key components and what's happening in this code:

- `navigatorKey`: Essential for keeping track of the Navigator state, allowing the router to perform navigation actions such as pushing and popping routes
- `_selectedDessert`: Maintains the state of the currently selected dessert; when a user selects a dessert from the list, this property is updated, which then influences the route displayed to the user

The preceding code snippet will not work once you apply it to your code directly, because there are some parts we need to cover to make it work.

Implementing error handling

To manage error states, such as navigating to a non-existent dessert or an invalid URL, we introduce a `show404` flag:

```
bool show404 = false;
```

Adapting the onPopPage method

Adapting this method is important to handle back navigation with the app:

```
onPopPage: (route, result) {
  if (!route.didPop(result)) {
    return false;
  }
  setState(() {
    _selectedDessert = null;
    show404 = false;
  });
  return true;
},
```

Implementing error handling with an UnknownScreen

To enhance user experience and manage navigation errors effectively, we introduce an `UnknownScreen` widget. This widget is displayed whenever users navigate to a route that doesn't exist within our app, providing a clear and friendly error message:

```
Scaffold(
  body: Column(
    children: [
      Text('404 NOT FOUND'),
      Text('The dessert you are looking for is eaten or it was never
here!'),
    ],
  ),
)
```

Update your navigation logic to include the `UnknownScreen` when the `show404` flag is `true`. This adjustment is made within the list of pages managed by your `Navigator`:

```
if (show404)
  const MaterialPage(key: ValueKey('UnknownPage'),
  child: UnknownScreen()
)
```

Reflecting navigation states in the URL

Next, let's override the `currentConfiguration` getter in `DessertRouterDelegate` to accurately reflect the app's current state in the URL. This involves returning a `DessertRoutePath.unknown()` when `show404` is `true`, a `DessertRoutePath.home()` when no dessert is selected, and a `DessertRoutePath.details()` with the correct index when a dessert is selected:

```
@Override
DessertRoutePath get currentConfiguration {
  if (show404) {
    return DessertRoutePath.unknown();
  }
  return _selectedDessert == null
      ? DessertRoutePath.home()
      : DessertRoutePath.details(desserts.indexOf(_selectedDessert!));
}
```

And here's what utilizing the `Navigator` with `navigatorKey` looks like:

```
Navigator(
  key: navigatorKey,
  // Other properties or children can be added here
)
```

We have a `Navigator` widget in our app that uses a special key, `navigatorKey`, to keep track of the navigation history and state. This `Navigator` widget decides what screen or page to show based on the current app state:

- It starts with showing a list of desserts on the `DessertsListScreen`. Here, you can choose any dessert to see more details.

- If there's an error or the page can't be found (such as when a wrong URL is entered), it shows an `UnknownScreen` to let you know that something went wrong.

- When you select a dessert, it shows the `DessertDetailsScreen` to give you more info about the chosen dessert.

Now, our `DessertRouterDelegate` is more than just a simple class; it's a `ChangeNotifier`. This change allows us to use `notifyListeners` instead of `setState` for updating the app's state.

`notifyListeners` tells everyone listening (such as our Router widget) that something has changed, so they should look again and update what they're showing. This ensures that the app reacts properly. It updates the displayed page and the URL in the browser, keeping everything in sync.

This is part of moving from managing states within a widget to using broader app-level state management. It makes our app smarter about when to update and redraw screens, improving performance and user experience.

Exploring the navigation with ChangeNotifier in DessertRouterDelegate

With `DessertRouterDelegate` now extending `ChangeNotifier`, `onPopPage` can use `notifyListeners` instead of `setState`. This change ensures that when the `RouterDelegate` notifies its listeners, the `Router` widget is alerted to the updated `currentConfiguration` and rebuilds the Navigator accordingly.

Adapting to notifyListeners for state updates

Our `_handleDessertTapped` should be adapted to `notifyListeners` too:

```
void _handleDessertTapped(Dessert dessert) {
    _selectedDessert = dessert;
    notifyListeners();
  }
```

This method ensures that whenever a dessert is selected, the app knows to show details about that dessert.

Dynamic route handling with setNewRoutePath

The `setNewRoutePath` function is called by the `Router` when there's a change in the app's navigation route. It's an opportunity for the app to adjust its state based on the new route information. We put all that into practice in the following code block:

```
Future<void> setNewRoutePath(DessertRoutePath configuration) async {
  if (configuration.isUnknown) {
    _selectedDessert = null;
    show404 = true;
    return;
  }
  if (configuration.isDetailsPage) {
    if (configuration.id! < 0 || configuration.id! >= desserts.length)
  {
      show404 = true;
      return;
    }
    _selectedDessert = desserts[configuration.id!];
  } else {
    _selectedDessert = null;
  }
  show404 = false;
}
```

Here's how it works:

- If the route leads to an unknown destination, the app resets the selected dessert and flags it to show a 404-error page, indicating that the requested page couldn't be found.

- If the route points to a dessert's details page, the app checks whether the dessert ID from the route is valid. If it's not (such as if the ID is out of range), the app prepares to show a 404 page. Otherwise, it updates the selected dessert to match the ID from the route.

- If the route is for the home page, the app resets the selected dessert, as no specific dessert is being viewed.

Next, let's learn more about our route parser.

Understanding DessertRouteInformationParser

The `DessertRouteInformationParser` is a specialized class designed to help your Flutter app interpret and manage web URLs effectively. It ensures that navigation within your app is smooth and intuitive, whether users click on links or type URLs directly into their browsers. Let's break it down into two main concepts:

- **Parsing home and details routes**:

  - If the URL is simple, such as `/`, it means the visitor wants to go to the home page where all desserts are listed.

  - If the URL follows a pattern such as `/dessert/1`, it indicates that the visitor wants to see details about a specific dessert. The number at the end tells us which dessert they're interested in.

- **Handling unknown routes**: Sometimes, a visitor might try to go to a URL that doesn't match any of our defined paths such as `/dessert/abc`. When this happens, we treat it as an unknown route, like how a website shows a **404 Page Not Found** error.

Now, let's implement the class.

Implementing DessertRouteInformationParser

This class translates URLs into actionable instructions for your app, determining which page to display based on the URL structure:

```
class DessertRouteInformationParser extends
RouteInformationParser<DessertRoutePath> {
  @override
  Future<DessertRoutePath> parseRouteInformation(RouteInformation
routeInformation) async {
    final uri = routeInformation.uri;
    if (uri.pathSegments.isEmpty) {
      return DessertRoutePath.home();
    }
    if (uri.pathSegments.length == 2) {
      if (uri.pathSegments[0] != 'dessert') return DessertRoutePath.
unknown();
      var id = int.tryParse(uri.pathSegments[1]);
      if (id == null) return DessertRoutePath.unknown();
      return DessertRoutePath.details(id);
    }
    return DessertRoutePath.unknown();
  }
```

```
    @override
    RouteInformation? restoreRouteInformation(DessertRoutePath
    configuration) {
        if (configuration.isUnknown) {
            return RouteInformation(uri: Uri.parse('/404'));
        }
        if (configuration.isHomePage) {
            return RouteInformation(uri: Uri.parse('/'));
        }
        if (configuration.isDetailsPage) {
            return RouteInformation(uri: Uri.parse('/
    dessert/${configuration.id}'));
        }
        return null;
    }
}
```

The primary method, `parseRouteInformation`, begins by extracting the URI from the `RouteInformation` object to get the URL components. This step is essential because it breaks down the URL into manageable parts for further analysis. The method then checks whether the path segments are empty, which indicates the home page (`/`). If this condition is met, it returns a `DessertRoutePath.home()` object, directing the app to the home page where all desserts are listed.

For URLs with two path segments, the method first ensures that the initial segment is `dessert`. This check helps verify that the URL is intended for a dessert details page. If the first segment is not `dessert`, it returns `DessertRoutePath.unknown()`, signaling an invalid or unknown route. If the initial segment is correct, the method attempts to parse the second segment as an integer, representing the dessert ID. Successfully parsing this segment allows the method to return a `DessertRoutePath.details(id)` object, guiding the app to display details for the specified dessert. If parsing fails, it again returns `DessertRoutePath.unknown()`. Any other URL pattern that does not match these conditions is also treated as unknown, with the method returning `DessertRoutePath.unknown()`.

The `restoreRouteInformation` method complements this by converting a `DessertRoutePath` back into `RouteInformation`. If the path configuration indicates an unknown route, it generates a `RouteInformation` object with the `/404` URI, similar to a Page Not Found error, to handle invalid URLs properly. For the home page, it returns a `RouteInformation` object with the `/` URI, ensuring the browser's address bar reflects navigation to the home page. When dealing with details pages, it creates a `RouteInformation` object with the `/dessert/{id}` URI, accurately representing the dessert details being viewed. If the path configuration does not match any known patterns, it returns null, indicating no valid route.

By implementing these methods, the `DessertRouteInformationParser` ensures that the app can correctly interpret and respond to various URL structures, maintaining a smooth and intuitive user navigation experience.

Restoring route information

The class also helps in creating URLs based on the app's current state. This is important for web apps because it ensures that the URL in the browser's address bar matches what the user sees on their screen.

Here is some general advice for crafting URLs based on app state:

- **Unknown page**: When users navigate to a page that doesn't exist, the app uses `/404` in the URL, indicating an error similar to the traditional Page Not Found on websites.

- **Home page**: The `/` root URL signifies the home page, where users can see a list of all desserts.

- **Dessert details page**: Detailed views of desserts are represented by URLs following the `/dessert/1` pattern, with the number changing based on the specific dessert selected. This format makes it straightforward for users to share links or bookmark pages.

Now we need to integrate these pieces.

Integrating navigation with MaterialApp.router

Finally, to apply our routing logic, we switch from using `MaterialApp` to `MaterialApp.router`. This allows us to define our custom routing behavior using the `DessertRouteInformationParser` and a `RouterDelegate` (`DessertRouterDelegate`). We pass these components to `MaterialApp.router`, empowering our app to handle complex navigation scenarios and link-based navigation just like a web page:

```
MaterialApp.router(
    title: 'Candy Shop',
    routerDelegate: _routerDelegate,
    routeInformationParser: _routeInformationParser,
);
```

Since we have these components, now we can check how it works in the Chrome browser.

Navigating your app in Chrome

When you run your Flutter app in Chrome, you'll get to see the navigation in action, reflecting in the browser's URL bar as you move around the app. This is what you can do and see:

- **Viewing the list of desserts**: Initially, you'll land on the home page where all the desserts are listed. The URL will look something like this: `http://localhost:64577/`.

- **Selecting a dessert**: As you click on any dessert to see more details, watch the URL change to reflect your selection. For the first dessert, the URL will update to `http://localhost:64577/#/dessert/0`.

- **Manually editing the URL**: Try changing the dessert's index number in the URL to navigate directly to a different dessert. For example, updating the URL to `http://localhost:64577/#/dessert/2` will take you to the third dessert in the list.

- **Encountering an unknown route**: If you get curious and change the index to a number outside the range of available desserts, such as `http://localhost:64577/#/dessert/10000`, and press *Enter*, the app will direct you to an `UnknownScreen`. The URL will also reflect this by changing to `http://localhost:64577/#/404`, indicating that you've reached a non-existent page within the app.

This dynamic URL and manual navigation capability brings your Flutter app closer to the experience of navigating a traditional website, making use of deep linking and browser navigation features such as the back and forward buttons. If you ever feel lost or need to reference the complete setup, the `final_navigation_2_declarative/candy_store/lib/main.dart` file in *Chapter 5* has everything you need. It's a great resource to see the navigation system fully implemented and to try out these interactions yourself.

This step in developing your Flutter app not only enhances user experience but also familiarizes you with advanced navigation patterns, preparing you for building more complex and intuitive apps.

Next, we'll compare 1.0 with 2.0 and understand the differences.

Comparing Navigation 1.0 and 2.0

Understanding the differences between Navigation 2.0 and the imperative approach (Navigation 1.0) is important for selecting the right navigation strategy for your Flutter application:

- **Centralized navigation state**: Navigator 2.0 helps us manage our app's navigation state in a single place, making it easier to handle complex scenarios.

- **Conditional navigation**: Easily manage navigation based on state or user permissions using this. For example, redirecting an unauthenticated user to a login screen is simple with conditional navigation.

- **Deep linking**: Navigator 2.0 simplifies handling deep links by allowing the app to update its navigation state in response to incoming URLs. We will learn about deep linking shortly.

The imperative approach is action-driven. For simpler applications with straightforward navigation flows, the imperative approach can be more than sufficient and easier to implement.

Despite the advanced capabilities of Navigation 2.0, the imperative approach remains effective for certain scenarios:

- **Small to medium apps**: Apps with a linear navigation flow and fewer screens may find the imperative approach easier to manage and understand

- **Quick prototyping**: When speed is of the essence, and you need to prototype an app quickly, the imperative approach allows for rapid development without worrying about the navigation state

Navigator 2.0, with its declarative routing approach, represents a powerful tool for us and helps us have more dynamic and responsive navigation patterns. While there's a learning curve to effectively leverage its capabilities, the flexibility and control it offers make it a valuable addition to the app framework for building complex and user-friendly applications.

Imperative navigation, with its straightforward implementation and ease of use, continues to be a viable option for many projects, particularly those with simpler navigation needs. It offers a lower entry barrier for developers and facilitates quick results, making it suitable for applications where complex state management and deep linking are not critical priorities.

However, both navigation strategies have their rightful place in Flutter development. The choice between Navigation 2.0 and imperative navigation should be informed by your project's specific requirements.

Summary

You learned how to navigate between screens using the imperative approach of Navigator 1.0 in this chapter. This includes pushing and popping routes, which is fundamental for creating multi-page applications. Then we introduced Navigator 2.0, which provides a declarative way of handling navigation. This section explained how to manage complex navigation scenarios, such as dynamically changing routes and handling deep links, offering more control over the app's navigation state.

In our exploration of Flutter navigation, we built a solid foundation by understanding both Navigator 1.0 and 2.0. Knowing when and how to use each model based on your app's needs is crucial for effective navigation design. Transitioning from the imperative style of Navigator 1.0 to the declarative approach of Navigator 2.0 equipped us with the skills to handle more complex navigation patterns, thereby improving the scalability and maintainability of our code. Mastering deep linking and URL management with Navigator 2.0 prepared us to create Flutter web applications that behave like traditional web apps, improving the user experience with familiar navigation patterns. This knowledge also helps us understand navigation patterns for mobile applications more easily. Additionally, learning to handle unknown routes has taught us to manage user errors and unexpected navigation paths, ensuring a robust and user-friendly app.

In the next chapter, we will learn about the responsible repository pattern, which is how we will set our repositories and build data sources, explore how to retrieve data, explore caching, and more. It will help us to bridge the gap between our app data sources and business logic.

Part 3: Exploring Practical Design Patterns and Architecture Best Practices

In this part, we will continue filling our toolbox with useful design and architecture patterns and best practices. We will explore a variety of topics, starting with how to build an efficient data layer with repository patterns and various **dependency injection (DI)** practices. We will then look at multiple approaches to layered architecture, learn how to implement them, and reason about software design principles such as SOLID, DRY, and KISS. We will proceed by diving into the best practices for managing concurrency in Dart, and finally, learn how to efficiently tackle tasks that take us beyond Flutter, to the native platform layer.

This part includes the following chapters:

- *Chapter 6, The Responsible Repository Pattern*
- *Chapter 7, Implementing the Inversion of Control Principle*
- *Chapter 8, Ensuring Scalability and Maintainability with Layered Architecture*
- *Chapter 9, Mastering Concurrent Programming in Dart*
- *Chapter 10, A Bridge to the Native Side of Development*

6

The Responsible
Repository Pattern

In the ever-evolving landscape of mobile application development, achieving a clean and maintainable architecture is a constant pursuit. As developers, we strive to separate the concerns of the **user interface** (**UI**) state from the actual business logic that makes our applications functional. As we have seen in *Chapter 4*, we adopt state management solutions such as bloc, cubit, ViewModel, or Provider to efficiently manage the UI state, bringing clarity and order to our code base.

However, as we dive deeper into the complexities of building robust Flutter apps, we inevitably encounter a new set of challenges. The process of segregating the UI state from business logic can sometimes leave us grappling with dilemmas such as duplicating domain logic, wrestling with data sources, and battling inconsistent behavior within our applications. These issues can quickly spiral out of control, leading to code base chaos and maintenance nightmares.

In this chapter, we'll address and conquer these challenges head-on. We will introduce you to a powerful architectural pattern: the **repository pattern**. With the help of the repository pattern, we will learn how to improve our Flutter app's architecture, ensuring that it remains both scalable and maintainable.

By the end of this chapter, you will gain a comprehensive understanding of how the repository pattern can be applied to overcome these complex challenges. We will dive deep into its principles and practical applications, equipping you with the knowledge and tools necessary to build an architecture that not only stands the test of time but also thrives in the dynamic world of mobile app development.

In this chapter, we're going to cover the following main topics:

- Introducing the repository pattern
- Setting up our first repository
- Defining repository interfaces
- Building a data source

- Creating a local database using Hive
- Implementing repository data retrieval
- Integrating the repository with our bloc
- Caching strategy
- Understanding data synchronization

Technical requirements

In order to proceed with this chapter, you will need the following:

- Code from the previous chapter, which can be found here: `https://github.com/PacktPublishing/Flutter-Design-Patterns-and-Best-Practices/tree/master/CH05/final_navigation_1/candy_store`
- All of the code required for this chapter, which can be found here:
  - Start of chapter: `https://github.com/PacktPublishing/Flutter-Design-Patterns-and-Best-Practices/tree/master/CH06/initial/candy_store`
  - End of chapter: `https://github.com/PacktPublishing/Flutter-Design-Patterns-and-Best-Practices/tree/master/CH06/final/candy_store`

  You can review the step-by-step refactoring in the commit history of this folder.

You will need to add these libraries to the Candy Store app:

- `https://pub.dev/packages/hive`
- `https://pub.dev/packages/hive_flutter`

Introducing the repository pattern

As we dive deeper into the complexities of building robust Flutter applications, we encounter the need for a structured approach to handling data. In *Chapter 4*, we introduced the notion of the *Model*, the architecture layer that encapsulates data structures and business logic, defining how data is structured, validated, and manipulated within the application. This is where the repository pattern comes in. In essence, the repository pattern establishes a structured mechanism for interacting with data sources, providing a standardized interface for the *Model* to execute and oversee data access operations.

The repository pattern is a software design pattern that focuses on separating the application's business logic from the details of data access. It achieves this by introducing a centralized component known as a repository. This repository acts as an intermediary layer between your app's data sources (such as databases, APIs, and external services) and the rest of your application. It encapsulates the logic for fetching and storing data, shielding the rest of your code from the intricacies of data retrieval and persistence.

Key to understanding the repository pattern is its ability to do the following:

- **Organize code**: The repository pattern helps keep your code base organized. It centralizes data access and ensures that your business logic remains uncluttered by data-fetching code. This separation of concerns makes your code base more maintainable and easier to comprehend.

- **Abstract data sources**: By abstracting the data sources, you can switch between different data storage mechanisms (for example, from a local database to a remote API) without affecting the rest of your application. This flexibility is especially valuable when your app evolves and needs to adapt to changing requirements.

- **Simplify testing**: The repository pattern simplifies unit testing. You can create mock repositories to isolate and test your business logic independently of the actual data sources, leading to more reliable and efficient testing.

- **Ensure consistency**: It enforces a consistent way to access and manipulate data, reducing the chances of bugs caused by inconsistent data operations scattered throughout your code base.

In essence, the repository pattern helps you control the data complexity in your Flutter app and promotes a clean separation of concerns, leading to code that's easier to maintain and extend.

How the repository pattern works

Now that we've explored the numerous advantages of adopting the repository pattern, let's transition into the implementation of this pattern by diving into its key components. Understanding these components is crucial for implementing the repository pattern effectively in your Flutter applications:

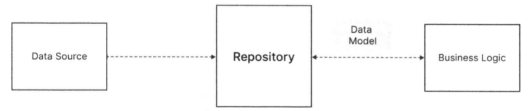

Figure 6.1 – Diagram of the three components of the repository pattern

Let's unpack this diagram:

- **Data sources**: Repositories interact with various data sources, which include local databases, remote APIs, and other external services, to fetch and store data. These are the sources where the repository retrieves or deposits data.

- **Repository layer**: Serving as the intermediary layer between data sources and business logic, the repository layer provides methods for your business logic to request data. It abstracts the complex details of data retrieval and persistence, simplifying your application's architecture.

- **Business logic**: The core logic of your application, distinct from data operations, interacts with repositories to access and manipulate data. This separation of concerns ensures that your business logic remains focused on its primary objectives, unburdened by the complexities of data retrieval and storage.

- **Data models**: Data retrieved from repositories is typically transformed into application-specific models before being utilized by the business logic. These models abstract the raw data into structured, meaningful representations that are tailored to your application's needs.

To illustrate its implementation, we'll use a practical example that is based on our Candy Store app.

Setting up our first repository

In this section, we will take our first steps into implementing the repository pattern for our app. In the context of the Candy Store app, and in general, the repository will serve as a bridge between our app's UI and the data sources. It will handle tasks such as retrieving a list of candies, adding new candies, and updating the candy stock. By doing so, it centralizes data access and ensures that our business logic remains unburdened by data-fetching code.

We will walk through the process of creating our first repository for the Candy Store app. We'll define its responsibilities, set up data sources, and leverage repository interfaces to build a clean and maintainable data access layer for our application.

Defining responsibilities

In this section, we'll take a practical approach to defining the responsibilities of our Candy Store app's repository within the context of our product list page. Instead of diving into theory, let's consider the specific needs of our app and how the repository can fulfill them:

- **Fetching candy data**: Our product list page must display a list of candies available for purchase. The repository's first responsibility is to fetch this data from our data sources. This includes querying a local database or making API requests to retrieve the latest candy information.

- **Adding new candies**: When our Candy Store app introduces new candies, users need to see these additions. The repository should provide a method for adding new candies to the data source, ensuring they appear in the product list.

- **Updating stock levels**: Candies get sold, and their stock levels change. The repository should handle updates to these stock levels, ensuring that users see accurate information about candy availability.

- **Error handling**: Real-world applications often face network issues or other errors. Our repository should implement error-handling mechanisms to gracefully handle and report errors to the UI layer, providing a smooth user experience.

- **Caching data**: To improve performance and reduce unnecessary network requests, our repository can implement a caching mechanism. It should store retrieved data locally and update it when necessary to provide a seamless user experience.

As we proceed with the implementation of our repository, we'll keep these practical responsibilities in mind. This hands-on approach will help us understand how the repository pattern can be a powerful tool in building maintainable and efficient Flutter applications such as our Candy Store app.

Creating our first repository

Now, let's get practical. We'll create our first repository in the context of our Candy Store app, keeping it simple and easy to follow. Our repository will be responsible for fetching a list of candies:

1. In your Flutter project, create a new Dart file for the repository. You can name it something like `product_repository.dart`. Inside this file, define a class for our repository. Let's call it `ProductRepository`.

lib/product_repository.dart

```
class ProductRepository {
  // Repository methods will go here
}
```

2. Inside this repository, add a method for fetching the list of candies. For simplicity, we'll return a list of strings representing candy names. In a real-world scenario, this method would interact with your chosen data source (for example, a database or API):

lib/simple_product_repository.dart

```
class SimpleProductRepository {
  List<String> fetchProducts() {
    // In a real app, this method would fetch data from a data
    source.
    return [
      'Chocolate Bar',
      'Gummy Bears',
      'Jelly Beans',
      'Lollipop',
      'Caramel Chew',
    ];
```

```
        }
    }
```

3. Now, let's use this repository in our app. For instance, you can integrate it into a Flutter widget, such as a `ListView`, to display the list of candies. Here's a simplified example:

lib/example_product_repository_with_screen.dart

```dart
import 'package:flutter/material.dart';

final SimpleProductRepository productRepository =
SimpleProductRepository();

class ProductListScreen extends StatefulWidget {
  const ProductListScreen({super.key});

  @override
  State<ProductListScreen> createState() => _ProductListScreenState();
}

class _ProductListScreenState extends State<ProductListScreen> {
  List<String> _products = [];

  @override
  void initState() {
    super.initState();
    _products = productRepository.fetchProducts();
  }

  @override
  Widget build(BuildContext context) {
    return Scaffold(
      appBar: AppBar(
        title: const Text('Candy Store'),
      ),
      body: ListView.builder(
        itemCount: _products.length,
        itemBuilder: (context, index) {
          return ListTile(
            title: Text(_products[index]),
          );
        },
      ),
```

```
        );
    }
}
```

In this example, we've created a `ProductListScreen` widget that uses our `SimpleProductRepository` to fetch and display a list of candies. Remember that this is a simplified illustration to help you grasp the basic concept.

As you progress in your Flutter app development journey, you can expand this repository to include more complex data retrieval logic, handle data sources such as databases and APIs, and apply advanced repository patterns for caching and offline capabilities. The key takeaway here is that repositories serve as an essential bridge between your app's data and business logic, promoting clean and maintainable code.

Defining repository interfaces

Now that we've gained a solid understanding of how to create and define the responsibilities of a repository, let's explore how we can leverage repository interfaces to enhance the flexibility and testability of our code.

In Flutter, using interfaces isn't a strict requirement, but it can significantly improve your code's maintainability, flexibility, and testability. Repository interfaces act as blueprints for repositories, defining the contract that each repository must adhere to. While implementing the repository, following this contract ensures that the repository behaves consistently across your app.

Interfaces in Dart

If you're coming to the Dart language from languages such as Java, you might be surprised that interfaces and classes can behave differently from what you expect. For example, in Java, if you defined an entity with a keyword `interface`, then in any class definition you can only `implement` that interface, and `extend` another class, hence the interfaces are **explicit**.

On the other hand, in Dart, all classes implicitly define an interface by default. This means that if you define a class, another class can either **extend** it—thereby inheriting all its behavior—or **implement** it, inheriting only the contract without the functionality.

This has slightly changed with Dart language version 3.0, as new modifiers became available: `interface`, `base`, `final`, and `sealed`. These modifiers add various properties and can also be combined. You can read more about them in the official documentation (as of 28.07.2024 `https://dart.dev/language/class-modifiers`).

That said, while we may still use explicit interfaces, it doesn't mean you have to do so in your own code, as use cases can be different. In this chapter and throughout this book, when we refer to an interface, we mean the general concept of an interface as a contract, not exclusively an explicit interface in the Dart language.

To create a repository interface, you'll define a set of method signatures that outline the operations your repository will support. For our `CandyRepository`, these methods might include the following:

- `Future<List<Product>> fetchProducts()`: Retrieves a list of candies from the data source
- `Future<Product> fetchProductById(int id)`: Fetches a specific candy by its unique identifier
- `Future<void> updateProduct(Product product)`: Allows for updating candy data

By creating these interface methods, you establish a clear contract for what developers can expect from any repository implementing this interface.

Why use repository interfaces?

There are a number of benefits to using repository interfaces:

- **Flexibility**: Repository interfaces make it easy to swap out different repository implementations without altering the rest of your code. For instance, you can switch between using a local database or an API-based repository by simply changing the repository instance you use throughout your app.
- **Testability**: With repository interfaces, you can create mock repositories that simulate various data scenarios. This is invaluable for unit testing your business logic independently of the actual data sources.
- **Consistency**: Interfaces enforce a consistent structure for repositories, helping developers understand how to use them. This consistency improves code readability and maintainability.

Implementing repository interfaces

Now that we appreciate the benefits of repository interfaces, let's see how to implement them for our `ProductRepository`. Here's a practical example.

First, we create our repository interface:

lib/product_repository.dart

```
abstract interface class ProductRepository {
  Future<List<Product>> fetchProduct();
  Future<Product> fetchProductById(int id);
  Future<void> updateProduct(Product product);
}
```

Secondly we implement this interface and create the actual repository.

```
class AppProductRepository implements ProductRepository {
  @override
  Future<List<Product>> fetchProducts() {
    // Fetch candies from the data source
    // (e.g., local database or API)
    // Transform the raw data if necessary
    // Return the list of candies
  }

  @override
  Future<Product> fetchProductById(int id) {
    // Fetch a specific Product by ID
    // Transform the raw data if necessary
    // Return the Product object
  }

  @override
  Future<void> updateProduct(Product product) {
    // Update product data in the data source
  }
}
```

In this example, we've created the `ProductRepository` interface with the specified methods. Then, we implemented these methods in `AppProductRepository`, which is the actual repository used in our app. This separation allows us to switch implementations or create mock repositories for testing while adhering to the defined contract.

The "I" prefix in the interface name

Depending on your background, you might be used to prefixing interface names with an "I", such as `IProductRepository`, or suffixing implementations with "Impl" (which stands for "implementation"), such as `ProductRepositoryImpl`. These practices are usually team conventions, influenced by the framework of choice or personal preference. While it's important to stick to what works best for your team, we will avoid these prefixes and suffixes in this scenario.

Whether an entity is an interface or a class is often considered an "implementation detail." This means it doesn't affect the behavior of the consumer class and is generally extra information, the explicitness of which can lead to tight coupling. Therefore, we will name our interfaces and classes based solely on their purpose, not their implementation details.

By implementing repository interfaces, we've made our code more flexible, testable, and maintainable, setting the stage for a robust architecture in our Candy Store app.

Moving forward, let's continue with building data sources - important entities that power our repositories.

Building a data source

Understanding data sources is crucial as they are the providers of data for our repositories, which are the core of our Candy Store app.

In Flutter, data sources serve as the gateways to your app's data, whether it's stored locally, fetched from remote APIs, or obtained from external services. They provide raw data that repositories then transform into usable information for your app's business logic and UI.

The various types of data sources include the following:

- **Local data sources**: These include local databases, such as SQLite, Drift, and Hive, as well as shared preferences. They are perfect for storing app data that needs to persist across app sessions.

- **Remote data sources**: Remote APIs, web services, and other online data providers fall under this category. These sources are essential for retrieving data from remote servers, such as fetching candy information from an online candy database.

- **External services**: These include external platforms and services such as Firebase, which offer real-time databases and authentication. Integrating these services can enhance your app's functionality and user experience.

But how do data sources and repository interfaces interact with one another?

Repository interfaces and data sources

Remember our `ProductRepository` interface we introduced earlier? It plays a vital role here. While data sources do not necessarily require interfaces, repositories' use of interfaces facilitates the decoupling of data source implementation from business logic. When you implement a repository, you'll specify which data source it should use by creating the corresponding method implementations.

In the next pages, we'll dive into creating a local database as one of our data sources and setting up API services for fetching candy data.

Setting up remote data sources

In the Candy Store app, we want to retrieve the product information from our servers where we have the most up-to-date availability of our candies. Let's dive into how to set up API services as data sources for our app.

To interact with our remote server, we'll create an `ApiService` class that handles HTTP requests for candy data. We can use popular HTTP packages such as `http` or `dio` to make network calls. Here, we'll use the `http` package as an example:

lib/api_service.dart

```
import 'dart:convert';
import 'package:http/http.dart' as http;
import 'product.dart';
class ApiService {
  final String _baseUrl = 'https://api.example.com/candystore';


  Future<List<Product>> fetchProducts() async {
    final response = await http.get(Uri.parse('$_baseUrl/products'));
    if (response.statusCode == 200) {
      final List<dynamic> productData = json.decode(response.body);
      return productData.map((json) => Product.fromJson(json)).
        toList();
    } else {
      throw Exception('Failed to load candies');
    }
  }
}
```

In the `ApiService` class, we've defined methods to fetch candy data from the API endpoint and convert the response into a list of `Product` objects.

Now that we have our `ApiService` class, let's integrate it into our Candy Store app. For example, to fetch a list of candies from the API, we could write code such as the following:

```
final apiService = ApiService();

try {
  final products = await apiService.fetchProducts();
  // Handle fetched candies
} catch (e) {
  // Handle error
}
```

To abstract our API service as a data source, we'll create a `NetworkProductRepository` class that provides methods to interact with the API for retrieving candy data. It will implement our `ProductRepository` interface, and we can use our implementation in any place that expects the interface:

lib/network_product_repository.dart

```dart
class NetworkProductRepository implements ProductRepository {
  NetworkProductRepository(this._apiService);

  final ApiService _apiService;

  @override
  Future<List<Product>> fetchProducts() {
    return _apiService.fetchProducts();
  }
}
```

In the previous NetworkProductRepository class, we can use the _apiService instance to perform **create, read, update, delete (CRUD)** operations on candy data.

Integrating NetworkProductRepository

Now that we have our NetworkProductRepository, let's integrate it into our Candy Store app:

1. For example, if we want to add a new candy to our remote API, we could write code similar to the following in the addProduct method of NetworkProductRepository:

    ```dart
    final apiDataSource = ProductApiDataSource(ApiService());
    final product = Product(
        name: 'New Candy',
        description: 'A tasty new treat',
        price: 1.99,
        imageUrl: 'assets/new_candy.jpg',
      );
    await apiDataSource.addProduct(newProduct);
    ```

2. We could write the following to retrieve a list of candies in the getProducts method of NetworkProductRepository:

    ```dart
    final products = await apiDataSource.getProducts();
    ```

Keep in mind that this is just sample code to demonstrate potential use. We will soon see how to actually integrate it into our Candy Store app. By using the NetworkProductRepository, we've effectively abstracted our remote API, making it easy to switch to other data sources such as a local database in the future, without major changes to our app's code.

Creating a local database using Hive

In this section, we'll explore a popular local database solution for Flutter apps – **Hive**. Hive (https://pub.dev/packages/hive) is a lightweight and efficient NoSQL database that's well-suited to mobile applications due to its speed and simplicity.

> **What is a NoSQL database?**
>
> SQL databases, using Structured Query Language, organize data into tables with predefined schemas, ideal for complex queries and strict data integrity. NoSQL databases, diverse in structure, handle unstructured data with flexibility and scalability. Key differences lie in data models, schemas, and suitability for specific project needs.

Adding Hive to your project

Before we dive into creating our local database, make sure to add the Hive package to your Flutter project's dependencies. You can do this by adding the following lines to your pubspec.yaml file:

```
dependencies:
  hive: ^2.2.3
  hive_flutter: ^1.1.0
```

After adding the dependencies, run flutter pub get to fetch them.

To use Hive, you need to initialize it in your Flutter app. Typically, this is done in the main function before running the app. For example, this is how you could initialize it right in the main method. However, in our app, we will abstract it away, as we will soon see:

```
void main() async {
  await Hive.initFlutter();
  runApp(MyApp());
}
```

In Hive, data is stored in boxes. Each box can be thought of as a separate database table. For our Candy Store app, we'll create a candies box to store candy-related data. In order to do that, we will create a new service and call it HiveService:

lib/hive_service.dart

```
class HiveService {
  Future<void> initializeHive() async {
    await Hive.initFlutter();
    Hive.registerAdapter(ProductAdapter());
    await Hive.openBox<Product>('products');
```

```
    }

    Box<Product> getProductBox() {
      return Hive.box<Product>('products');
    }
```

In the preceding code, we created a `HiveService` to manage Hive initialization and access to the candies box. To serialize and deserialize candy objects, we will need to define a custom Hive `TypeAdapter` that we have named `ProductAdapter`:

lib/hive_service.dart

```
class ProductAdapter extends TypeAdapter<Product> {
  @override
  final typeId = 0;

  @override
  Product read(BinaryReader reader) {
    return Product(
      name: reader.read(),
      description: reader.read(),
      price: reader.read(),
      imageUrl: reader.read(),
    );
  }

  @override
  void write(BinaryWriter writer, Product obj) {
    writer.write(obj.name);
    writer.write(obj.description);
    writer.write(obj.price);
    writer.write(obj.imageUrl);
  }
}
```

We have overridden `read` and `write` methods, as well as `typeId`, so that Hive knows how to serialize our `Product` data model.

Now that we have our Hive box ready, we can easily store candy data. For example, we could write code such as the following:

```
final productBox = HiveService().getProductBox();

final product = Product(
```

```
    name: 'Chocolate Bar',
    description: 'Delicious milk chocolate',
    price: 2.99,
    imageUrl: 'assets/chocolate_bar.jpg',
  );
  productBox.add(product);
```

We have created an instance of `HiveService`, requested the `productBox` via the `getProductBox` method, and added a new `Product` to it by calling the `add` method of the `productBox`.

Retrieving data is just as straightforward:

```
final products = HiveService().getProductBox().values;
```

In order to do that, we need to just access the `values` field of the box.

Hive's simplicity and speed make it an excellent choice for local data storage in Flutter apps. With our product box in place, we're ready to store and retrieve candy information as we continue building our Candy Store app.

Now that we have set up Hive as our local database, let's integrate it as a data source into our Candy Store app.

Creating a local data source

To abstract our local database as a data source, we'll create a `LocalProductRepository` class that provides methods to interact with Hive for storing and retrieving candy data and implements our already familiar `ProductRepository`:

lib/local_product_repository.dart

```
class LocalProductRepository implements ProductRepository {
  LocalProductRepository(this._productBox);

  final Box<Product> _productBox;

  @override
  Future<List<Product>> fetchProducts() async {
    return _productBox.values.toList();
  }
}
```

In the previous `LocalProductRepository` class, we can use the `_productBox` instance to perform CRUD operations on candy data. For now, we only have the method to fetch products, but we can add all of the other necessary operations in the future.

Now that we have our `LocalProductRepository`, let's integrate it into our Candy Store app.

Repository interfaces

At this point, you might wonder how the repository pattern fits into this setup. Repository interfaces play a crucial role in abstracting data sources, allowing us to seamlessly switch between them.

In our Candy Store app, we defined a `ProductRepository` interface with methods such as `fetchProducts` that are implemented differently based on whether we're using local or API data sources. This abstraction allows us to maintain a consistent interface for data retrieval throughout the app. The following will remind you what it consists of:

lib/product_repository.dart

```
abstract interface class ProductRepository {
  Future<List<Product>> fetchProducts();
}
```

With the `ProductRepository` interface in place, we can create concrete implementations for both `NetworkProductRepository` and `LocalProductRepository`, ensuring a clear separation of concerns in our app's architecture.

With our API services set up and integrated into the app, we now have a comprehensive data management system for the Candy Store app. In the following sections, we'll explore how the repository pattern ties everything together, making data retrieval and management even more seamless and efficient.

Implementing repository data retrieval

With our data sources – both local and remote – in place, it's time to implement repository data retrieval. The repository pattern plays a pivotal role here, serving as the intermediary layer between our app's business logic and data sources.

Enhancing our product repository

Let's begin by modifying our `AppProductRepository`. The repository will abstract the underlying data sources, allowing us to fetch candy data without worrying about whether it's coming from a local database or an API service:

lib/app_product_repository.dart

```
class AppProductRepository implements ProductRepository {

  AppProductRepository({
```

```
      required NetworkProductRepository remoteDataSource,
      required LocalProductRepository localDataSource,
  })  : _remoteDataSource = remoteDataSource,
        _localDataSource = localDataSource;

  final NetworkProductRepository _remoteDataSource;
  final LocalProductRepository _localDataSource;

  @override
  Future<List<Product>> fetchProducts() async {
    // Retrieve candies from local data source
    final localProducts = await _localDataSource.fetchProducts();

    // Check if local data source has data
    if (localProducts.isNotEmpty) {
      return localProducts;
    } else {
      // If local data source is empty,
      // fetch from API and cache it locally
      final apiProducts = await _remoteDataSource.fetchProducts();
      await _localDataSource.cacheProducts(apiProducts);
      return apiProducts;
    }
  }
}
```

In this `AppProductRepository` class, we inject both the local and API data sources through its constructor. When the `fetchProducts` method is called, it first tries to retrieve candies from the local data source. If the local data source contains data, it returns that data. If the local data source is empty, it fetches data from the API data source, caches it locally, and then returns it.

Integrating the repository

Now, let's integrate our `AppProductRepository` into our Candy Store app. In your UI components or business logic, you can create an instance of the repository and use it to fetch candies. Potentially, your code could look like the following:

```
final productRepository = AppProductRepository(
  ProductHiveDataSource(), // Your local data source
  ProductApiDataSource(), // Your API data source
);


try {
```

```
    final products = await productRepository.fetchProducts();
    // Handle fetched candies
} catch (e) {
    // Handle error
}
```

By using the `AppProductRepository`, you abstract away the complexities of data retrieval and caching, making your code cleaner and more maintainable. On the other hand, if your app doesn't require caching logic just yet, you can substitute `AppProductRepository` for the more specialized version, such as `NetworkProductRepository`. Or if you're just developing the feature and don't even know yet what exactly the data source will be, you can use an `InMemoryProductRepository` implementation, and later on switch it to `FirebaseProductRepository`. As long as you're using the `ProductRepository` interface, switching between them will be painless. We will see how exactly this switching can take place in the next *Chapter 7*, where we will discuss how to provide dependencies in detail.

Implementing repository data retrieval simplifies how we access data in our Candy Store app. Whether it's local or remote data, the repository pattern ensures that our code remains organized and adaptable.

In the next sections, we'll explore more ways to enhance our app's architecture and functionality using this pattern.

Integrating the repository with our business logic

In our pursuit of a well-structured and efficient Flutter app, we're now at a critical point – integrating our repository into the UI. As you may recall from *Chapter 4*, we adopted the bloc pattern to manage our app's state. We'll now apply the power of our repository to the bloc pattern. This integration brings us one step closer to achieving a clean and maintainable architecture for our Candy Store app.

The `ProductsBloc` will serve as the bridge between our UI and the repository. This bloc will manage the state related to `ProductsPage` in our Candy Store app. To set this up, you'll need to create the `ProductsBloc`, which will have methods for retrieving and updating product information.

Let's walk through the steps:

1.   Create the `ProductsBloc`. Start by creating a new bloc class, `ProductsBloc`. This class will extend the `Bloc` class and manage the state for product-related data:

lib/products_bloc.dart

```
import 'package:flutter_bloc/flutter_bloc.dart';

class ProductsBloc extends Bloc<ProductsEvent, ProductsState> {
  ProductsBloc() : super(ProductsState());
}
```

Later, we will add the methods to retrieve and update product data here.

2. We will also need to implement the `ProductsState` and `ProductsEvent`. The state will contain the products retrieved from the repository:

lib/products_bloc_state.dart

```
class ProductsState extends Equatable {
  final List<ProductListItem> items;
  final DelayedResult<void> loadingResult;

  const ProductsState({
    this.loadingResult = const DelayedResult.idle(),
    this.items = const [],
  });

  @override
  List<Object?> get props => [items, loadingResult];

  ProductsState copyWith({
    List<ProductListItem>? items,
    DelayedResult<void>? loadingResult,
  }) {
    return ProductsState(
      items: items ?? this.items,
      loadingResult: loadingResult ?? this.loadingResult,
    );
  }
}
```

lib/products_bloc_event.dart

```
sealed class ProductsEvent extends Equatable {
  const ProductsEvent();

  @override
  List<Object?> get props => [];
}

final class FetchProducts extends ProductsEvent {
  const FetchProducts();
}
```

3. Connect the repository. To fetch and manage product data, the `ProductsBloc` will communicate
    with the `AppProductRepository` we've set up. Note that currently, the services instances
    `apiService` and `hiveService` are imported as global fields from `main.dart`. We will
    fix this in *Chapter 7*:

lib/products_bloc.dart

```dart
import 'main.dart';

class ProductsBloc extends Bloc<ProductsEvent, ProductsState> {
  late final _productRepository = AppProductRepository(
    remoteDataSource: NetworkProductRepository(apiService),
    localDataSource: LocalProductRepository(
      hiveService.getProductBox(),
    ),
  );

  ProductsBloc() : super(const ProductsState()) {
    on<FetchProducts>(__onFetchProducts);
  }

  Future<void> __onFetchProducts(
    FetchProducts event,
    Emitter<ProductsState> emit,
  ) async {
    try {
      emit(
        state.copyWith(
          loadingResult: const DelayedResult.inProgress(),
        ),
      );

      final products = await _productRepository.fetchProducts();

      emit(
        state.copyWith(
          items: products
              .map(
                (p) => ProductListItem(
                  id: p.id,
                  name: p.name,
                  description: p.description,
                  price: p.price,
```

```
                          imageUrl: p.imageUrl,
                        ),
                      )
                      .toList(),
                ),
            );

        emit(
          state.copyWith(
            loadingResult: const DelayedResult.idle(),
          ),
        );
      } on Exception catch (ex) {
        emit(
          state.copyWith(
            loadingResult: DelayedResult.fromError(ex),
          ),
        );
      }
    }
  }
}
```

4. Finally, we will connect the `ProductsBloc` with our previously created `ProductsPage`. For that, we will make use of `BlocProvider` to provide the bloc instance to the UI and be able to read the products from there:

lib/products_page.dart

```
class ProductsPage extends StatelessWidget {
  const ProductsPage({super.key});

  @override
  Widget build(BuildContext context) {
    return BlocProvider(
        create: (context) => ProductsBloc()
          ..add(
            FetchProducts(),
          ),
        child: _ProductsView());
  }
}

class _ProductsView extends StatelessWidget {
  @override
```

```
Widget build(BuildContext context) {
  final items = context.select(
    (ProductsBloc bloc) => bloc.state.items,
  );
  return Scaffold(
    appBar: AppBar(
      title: const Text('Products'),
    ),
    body: ListView.builder(
      padding: const EdgeInsets.symmetric(vertical: 16),
      itemCount: items.length,
      itemBuilder: (context, index) {
        final item = items[index];
        return ProductListItemView(item: item);
      },
    ),
  );
}
}
```

By following these practical steps, your repository pattern remains in harmony with your preferred state management solution and you can efficiently manage data in your Candy Store application.

Bridging the gap between foundational setup and advanced strategies, let's explore some of the caching strategies to master the repository pattern.

Caching strategies

In our journey to master the repository pattern, we've explored the basics of setting up repositories and data sources and integrating them into our Flutter app. Now, let's dive into an advanced topic: **caching** strategies.

Caching is a crucial aspect of modern mobile app development. It involves storing frequently accessed data in a local cache to reduce the need for repeated network requests. In our Candy Store app, caching can greatly improve performance by providing faster access to frequently viewed candy information.

Caching strategy

In the previous section, we learned how we can leverage local databases such as Hive to temporarily store the data retrieved from our remote APIs. To optimize caching further, consider implementing a caching strategy based on your app's specific requirements. For example, you can set a maximum cache size, implement cache eviction policies, or refresh cached data periodically to keep it up-to-date.

Caching is a powerful technique that enhances the performance of our Flutter app. By implementing caching strategies tailored to our Candy Store app's needs, we can provide users with a smoother and more responsive experience.

In the next section, we'll explore implementing offline mode, allowing users to access candy data even when they're not connected to the internet.

Implementing offline mode

As we continue to explore advanced repository patterns in our journey to master the repository pattern in Flutter, it's time to tackle a crucial feature: **offline mode**. Enabling offline access to our Candy Store app's data ensures that users can continue browsing even when they have limited or no internet connectivity.

Offline mode is not just a nice-to-have feature; it's often a necessity. Users may find themselves in areas with poor network coverage or want to browse our candy store while on an airplane or in other offline scenarios. Implementing offline mode can significantly enhance the user experience and increase engagement with our app.

To implement offline mode effectively, we'll need to devise strategies for storing and retrieving data locally on the user's device. Here's a high-level overview of the steps involved:

- **Data synchronization**: Implement a mechanism to synchronize data between the local storage (on the user's device) and the remote server (our API). This synchronization should occur whenever the device has internet connectivity.

- **Local database**: Utilize a local database, such as Hive or SQLite, to store essential data locally on the user's device. This local data should mirror the data available on the server.

- **Offline access**: Modify our app's data retrieval logic to prioritize local data when the device is offline. When online, the app should fetch and update data from the remote server.

- **User feedback**: Provide clear feedback to users about their offline status and the limitations of offline mode. For example, users may see a message such as "You are currently offline. Some data may be outdated."

Implementing offline mode using Hive or similar local databases can significantly enhance the reliability and usability of our Candy Store app. Users will appreciate being able to browse and shop for candies even when they are not connected to the internet.

In the next section, we'll explore strategies for synchronizing data sources to ensure that our local and remote data stays in sync.

Understanding data synchronization

As we near the conclusion of our journey through the repository pattern, it's essential to address a critical aspect of managing data in a Flutter app: **synchronizing data sources**. In this section, we'll explore the concept of data synchronization and how it can be implemented within our Candy Store app.

In many real-world applications, data doesn't reside solely in one place. Instead, it's distributed across multiple data sources, such as local databases and remote APIs. Keeping these data sources synchronized ensures that the information presented to users is consistent and up-to-date, regardless of where they access it from.

For our Candy Store app, this means making sure that the candy data stored in our local database matches the data on our remote server. This synchronization process involves periodic updates and conflict resolution to handle scenarios where changes might occur in both locations simultaneously.

Implementing data synchronization

To achieve data synchronization, we can follow these steps:

1. **Retrieve remote data**: Periodically fetch data from our remote API to check for updates. This can be done using packages such as `http` for making network requests.

2. **Compare local and remote data**: Compare the data retrieved from the remote API with the data stored in our local database. Identify any differences or updates.

3. **Resolve conflicts**: If there are conflicting changes between the local and remote data (for example, if a candy's name was updated on both sides), implement a conflict resolution strategy. This might involve user input or automated conflict resolution rules.

4. **Update local data**: Apply the updates from the remote data to the local database, ensuring that it matches the latest information from the server.

Automating synchronization

Automating the synchronization process is crucial to maintaining data consistency without user intervention. You can use background tasks or scheduled jobs to periodically trigger the synchronization process, ensuring that the data stays up-to-date.

Data synchronization is a vital aspect of modern app development, especially when dealing with distributed data sources. Implementing synchronization logic within your repository pattern ensures that your Flutter app maintains data consistency and provides a seamless user experience.

Now, let's reinforce our understanding of the repository pattern by creating another relevant repository for our app – the `CartRepository`.

Refactoring the CartModel

Before we dive into *Chapter 7*, let's use everything we've learned in this chapter to refactor our
`CartModel` to align it with the repository pattern:

1. First, we will create an interface called `CartRepository` that includes all the public methods
    and getters from the `CartModel`. It will look like the following:

lib//cart_repository.dart

```
abstract interface class CartRepository {
  Stream<CartInfo> get cartInfoStream;
  Future<CartInfo> get cartInfoFuture;
  Future<void> addToCart(ProductListItem item);
  Future<void> removeFromCart(CartListItem item);
}
```

So, we have now extracted everything public from the `CartModel` into the
abstract `CartRepository`.

2. Next, we need to refactor our `CartModel`. Firstly, it should implement the `CartRepository`.
    Secondly, we need to give it a name. A suitable name for our current implementation of this
    interface could be `InMemoryCartRepository`, indicating that the repository's state is
    stored in the application's memory and is short-lived. It exists only as long as our app remains
    in memory. Once the app is terminated, the state of this repository is also removed. The code
    will look like the following:

lib/cart_model.dart

```
class CartModel {
  CartModel._internal();

  static final CartModel _instance = CartModel._internal();

  static CartModel get instance => _instance;

  // code related to cart management
}
```

Afterwards, it will be:

lib/in_memory_cart_repository.dart

```
class InMemoryCartRepository implements CartRepository {

    // code related to cart management
}
```

In addition to renaming our class, notice that we have also removed the `CartModel.instance` getter.

Instead of accessing our class globally, in *Chapter 7*, we will learn how to access the repositories in other parts of our code base.

Summary

This chapter has provided you with a wealth of knowledge and practical insights about the repository pattern. As we conclude, let's revisit the essential lessons you've learned and consider the next steps in your Flutter development adventure.

You've grasped the fundamental essence of the repository pattern, a design architecture that separates data access concerns from the rest of your Flutter application. By doing so, it brings structure, organization, and maintainability to your code base. We explored how to create repositories directly within your Flutter projects, without the need for interfaces. This hands-on approach simplifies data management while preserving flexibility.

We also saw how repositories are vital intermediaries, bridging the gap between your app's data sources and its business logic. Their functions include data retrieval, storage, and synchronization, making your code modular and maintainable.

You've explored various data sources, from local databases using Hive to remote APIs. Data synchronization is crucial to maintaining consistency and has been highlighted as a key practice when dealing with distributed data sources.

As you continue expanding your Flutter skills, remember that the repository pattern is just one tool in your developer toolkit. While it's a powerful pattern for data management, there are other architectural patterns and best practices to explore, each with its own strengths and use cases. Ensure you choose the right tool for your specific project requirements.

In the upcoming *Chapter 7*, we will explore methods to supply our repositories to various components such as blocs, view models, notifiers, providers, and so on. We will explore new patterns such as dependency injection and service locator, gaining insights into their practical applications, benefits, and potential limitations to enrich our understanding of software design.

7

Implementing the Inversion of Control Principle

In *Chapter 6*, we started working on our *Model* layer by introducing the repository pattern and exploring its various benefits. One of the main themes of patterns in this book is decoupling responsibilities, which leads to improved flexibility, scalability, and maintainability. The first crucial step in achieving this is extracting all the logic of the *Model* layer into repositories. The next logical step is to follow the **inversion of control** (**IoC**) principle in order to design a good app architecture.

In this chapter, we will learn how to provide our repositories to our blocs (view models, notifiers, providers, and so on) in a way that allows us to easily replace these dependencies, control their instantiation, and seamlessly test our code in the future. We will discuss the rationale behind patterns such as **dependency injection** (**DI**) and **service locator** (**SL**), and explore their practical implementation along with their benefits and limitations.

The main topics covered in this chapter are as follows:

- Decoupling dependency creation from use
- Implementing the DI pattern using RepositoryProvider
- Implementing the SL pattern using get_it

Technical requirements

In order to proceed with this chapter, you will need the following:

- Code from the previous chapter, which can be found here: `https://github.com/PacktPublishing/Flutter-Design-Patterns-and-Best-Practices/tree/master/CH06/final/candy_store`.
- Libraries from `pub.dev` that we will connect to our application: `flutter_bloc` (already added in *Chapter 4*) (`https://pub.dev/packages/flutter_bloc`) and `get_it` (`https://pub.dev/packages/get_it`)

- You will find all of the code required for this chapter here:

  - **Start of the chapter**: `https://github.com/PacktPublishing/Flutter-Design-Patterns-and-Best-Practices/tree/master/CH07/initial/candy_store`.

  - **End of the chapter**: `https://github.com/PacktPublishing/Flutter-Design-Patterns-and-Best-Practices/tree/master/CH07/final/candy_store`. You can review the step-by-step refactoring in the commit history of this branch. The alternative implementation using `get_it` can be found here: `https://github.com/PacktPublishing/Flutter-Design-Patterns-and-Best-Practices/tree/master/CH07/final_extra/candy_store`.

Decoupling dependency creation from usage

The practice of decoupling dependencies is not specific to Flutter. There is a high chance that you have already encountered the concept of IoC, as well as specific patterns such as DI and SL. If you have, this chapter will highlight some Flutter-specific approaches, along with their pros and cons. If you haven't, that's not a problem. We will discuss them in detail now. However, before we delve into the terminology and provide a solution, we need to identify the problem. Let's revisit `CartModel`, which we created in *Chapter 4*. In *Chapter 6*, we refactored `CartModel` to adhere to the repository pattern by renaming it to `InMemoryCartRepository` and implementing the abstract `CartRepository`. Besides doing this, we have also removed the static constructor. Now, let's zoom in to understand why we did that.

Identifying the singleton pattern

The previous instantiation of `CartModel` used a very specific approach and looked like this:

lib/cart_model.dart

```
class CartModel {
  CartModel._internal(); // 1

  static final CartModel _instance = CartModel._internal(); // 2

  static CartModel get instance => _instance; // 3

  // code related to cart management
}
```

And here's the `cart_bloc`.

---

lib/cart_bloc.dart

```
class CartBloc extends Bloc<CartEvent, CartState> {
  final CartModel _cartModel = CartModel.instance; // 4
}
```

Let's recap what's going on here:

1. We create a private named constructor called `_internal`. This prevents callers from creating instances of the `CartModel` class.

2. We create a static final field called `_instance` of the `CartModel` type, which initializes an instance of `CartModel` using the private constructor mentioned previously. This field is also private.

3. Finally, we expose `_instance` through a public static getter called `instance`.

4. To access `CartModel`, we need to use this getter in `CartBloc` via `CartModel.instance`. This is the only way to access an instance of the `CartModel` class. This pattern is known as a **singleton pattern**.

A **singleton** means that only one instance of a class exists. Whenever we access our `CartModel` using the `instance` getter, we get the same instance. This allows us to call it in every bloc that needs it without worrying about its internal state. It will be shared and consistent everywhere. While this may sound useful, you may also hear that a singleton is considered an **antipattern**, which means it's not recommended. But why?

Well, in the way we have implemented this pattern, there are several limitations:

*   **Global state**: Introducing a global state, as done with `CartModel.instance`, can lead to inconsistency due to hard-to-track state changes from various parts of the application.

*   **Inflexible design**: Accessing the instance directly in consumer classes such as `CartBloc` creates a tight dependency. This makes it difficult to write automated tests and update dependencies without modifying all consumer classes. We will talk more about automated testing in *Chapter 11*.

*   **Tight coupling**: Since the classes are tightly coupled, the consumer classes have unnecessary awareness of the implementation details of their dependencies. Decoupling our code is important for flexibility, scalability, and maintainability.

So, is there a way to maintain the benefits of the singleton pattern while avoiding these limitations?

Introducing the IoC principle

The problem of creating relationships between elements of a software program has been a challenge since the early days of programming. One solution, or rather a principle to follow when handling dependencies, is the **IoC principle**. Its history dates back to the 80s, as the composition of different parts of a program was a relevant issue when programs became more complex.

IoC is an abstract notion that can have various meanings depending on the context and framework being used. In this context, we specifically refer to dependencies. The main idea of this principle is to separate the creation and use of dependencies, for the same reasons discussed when talking about the singleton pattern.

Martin Fowler, a renowned software engineer, wrote a prominent article back in 2004 that is still referenced today when discussing dependency management (`https://martinfowler.com/articles/injection.html`). We will align our terminology with this article. Fowler argues that "inversion of control" is a generic term and suggests focusing on specific implementations of this principle, such as DI and SL. This is exactly what we will explore in the following sections.

As you may recall from our refactoring exercise in *Chapter 6*, in addition to renaming our class, we also removed all of the singleton pattern logic. As a result, our code no longer compiles because the `CartModel.instance` getter does not exist! To address this issue, let's fix it by implementing the DI pattern.

Implementing the DI pattern via RepositoryProvider

The idea behind the DI pattern is that instead of creating or accessing dependencies internally, as we did in `CartBloc`, the dependencies of the consumer class are passed and created during runtime. This decouples the creation of the dependency from its use. In our case, `CartBloc` doesn't need to know where the dependency comes from, as long as it satisfies the contract. The question then becomes the following: where and how should we inject the dependency?

Injecting dependencies via a constructor

One convenient way to ensure the satisfaction of dependencies is to pass them to the constructor. By doing so, it becomes impossible to create an instance of this object without providing the necessary dependencies. Furthermore, this responsibility is delegated to the instantiating class, relieving the consumer class of this burden. Therefore, implementing this in `CartBloc` would look like the following code blocks.

Here is what the code block would look like before:

lib/cart_bloc.dart

```
class CartBloc extends Bloc<CartEvent, CartState> {
  final CartModel _cartModel = CartModel.instance; // 1

  CartBloc(); // 2

  // Code related to cart management from the previous chapters
}
```

Here is what the code block would look like after:

lib/cart_bloc.dart

```
class CartBloc extends Bloc<CartEvent, CartState> {
  final CartRepository _cartRepository; // 3

CartBloc({
    required CartRepository cartRepository,
  }) : _cartRepository = cartRepository, // 4

  // Code related to cart management from the previous chapters
}
```

So far, we have completed the following steps:

1. Previously, we manually instantiated the _cartModel field of the CartModel type in CartBloc using the CartModel.instance singleton access.

2. The constructor of CartBloc was empty.

3. Now, instead of using CartModel, we have a field of the CartRepository type. It is important to note that the type is the CartRepository interface, not the specific InMemoryCartRepository implementation that we created in the previous part. Additionally, we do not initialize it at all!

4. Since the initialization has been moved to the constructor, we now require CartRepository to be passed as a parameter when creating an instance of CartBloc. We have no knowledge of how this repository is created or whether it uses a singleton pattern. As long as it adheres to the CartRepository contract, we are satisfied. The usage and access to methods of this repository remain unchanged.

Now, we have a new problem: how do we create an instance of CartBloc?

Providing dependencies via RepositoryProvider

The good news is that we have already worked with this idea. In *Chapter 3*, we learned a lot about `InheritedWidget` and how it is used to provide various dependencies, such as `ViewModels`, to be accessed from the widget tree through `BuildContext`. Just to refresh your memory, we had `CartProvider`, which extended `InheritedWidget` and wrapped `MaterialApp` at the root level of our application. It looked like this:

lib/main.dart

```
void main() {
  runApp(
    CartProvider(
      cartNotifier: CartNotifier(),
      child: MaterialApp(
        title: 'Candy shop',
        theme: ThemeData(
          primarySwatch: Colors.lime,
        ),
        home: const MainPage(),
      ),
    ),
  );
}
```

We provided an instance of `CartNotifier` through `CartProvider` by adding it to the widget tree. Then, wherever it was needed, we accessed it using the context, such as in `ProductListItemView` like this:

lib/product_list_item_view.dart

```
class ProductListItemView extends StatelessWidget {

  @override
  Widget build(BuildContext context) {
    final cartNotifier = CartProvider.of(context);
    // read & submit data to/from cartNotifier
  }
}
```

This was made possible thanks to how `InheritedWidget` works. However, since we're using `flutter_bloc` for our state management, it provides a very similar solution out of the box. Instead of creating our custom providers that extend from `InheritedWidget`, we can use the

RepositoryProvider<T> class, which handles the "provider" part for us! It relies on the provider library (https://pub.dev/packages/provider) under the hood, which works in a similar way as InheritedWidget for this scenario, but also offers additional features that we don't need to consider at the moment. So, going back to our feature, our main function can now look like this:

lib/main.dart

```
void main() {
  runApp(
    RepositoryProvider<CartRepository>( // 1
      create: (_) => InMemoryCartRepository(), // 2
      child: MaterialApp(
        title: 'Candy shop',
        theme: ThemeData(
          primarySwatch: Colors.lime,
        ),
        home: MainPage.withBloc(),
      ),
    ),
  );
}
```

A very important point to note here is the following:

1. We create RepositoryProvider of the CartRepository type, which is an interface.

2. In the create method, we create a specific implementation, in this case, InMemoryCartRepository. If we ever need to change InMemoryCartRepository to something else, such as NetworkCartRepository or TestCartRepository, we can do so here without making any changes in the actual consumer classes! This is where the true beauty of decoupling, introduced by the Repository and DI patterns, shines.

> **Good to know**
>
> It is almost certain that in the scope of your application, you will need to provide more than one repository. While you can wrap one RepositoryProvider into another almost indefinitely, there is a more convenient way to do that – use MultiRepositoryProvider, which accepts a list of RepositoryProvider as an argument.

Finally, we need to connect our elements in one last place: in the actual MainPage widget that creates CartBloc. In *Chapter 4*, we have already seen how to use BlocProvider to provide a bloc to a portion of our widget tree. Now, we will use the same BlocProvider to pass the CartRepository dependency to CartBloc. We can do it as follows:

lib/main_page.dart

```
BlocProvider<CartBloc>(
    create: (context) => CartBloc(
      cartRepository: context.read(), // 1
    )..add(const Load()),
    child: const MainPage(),
  );
```

Notice how, on line 1, we read the instance of `CartRepository` from `context` using the `read` extension method. We have already done this in previous chapters. The `read` extension method from the `provider` package looks up the widget tree to find the dependency of the required type. If it doesn't find one, it throws an exception. In our case, we provide this dependency via a helper class called `RepositoryProvider`. It's important to note that this provider is "lazy" by default. It means that it only creates the actual instance of `InMemoryCartRepository` when it's explicitly requested for the first time. You can change this behavior by modifying the `lazy` parameter in the `RepositoryProvider` constructor. Since we have used `RepositoryProvider` at the top of our widget tree, in practice, it will be a singleton because the instance returned by `context.read` is the same. None of our blocs are aware of this implementation detail, which makes the whole architecture more flexible and decoupled.

However, one peculiarity of this approach, which can be both a pro and a con depending on your needs, is that all of our repositories and services are injected into the widget tree, potentially making it very large. It also means that they are bound to the same life cycle. Additionally, reading dependencies from the context in the *Model* layers is not something we should do because the context belongs to the *View* layer. We will discuss these layers and dependency scoping in detail in *Chapter 8*. Now, let's review an alternative approach as a pattern and as a tool. We will see how to decouple our widget tree and the DI mechanism using the SL pattern with the `get_it` library.

Implementing SL pattern via get_it

The SL pattern involves delegating dependency creation to a central registry, known as the *service locator*. Any consumer can then retrieve any dependency they need from this central registry, without needing to know how it is created. The main difference from DI is that consumers are typically aware of the SL and are tightly coupled to it. We will provide a more detailed comparison and discuss when to choose each pattern at the end of this chapter. For now, let's see how we can implement the SL pattern to gain a better understanding.

A popular library for implementing the SL pattern in Flutter is `get_it`. To use it, we need to add it to our `pubspec.yaml` file. Then, similar to `RepositoryProvider`, we need to register our dependencies using the `GetIt.instance` object. The syntax is similar to what we have seen with `CartModel.instance`. `GetIt` serves as a service registry where we register the dependencies we will need. This registry itself is a singleton, allowing us to access it whenever and wherever. Unlike `RepositoryProvider`, the type of our dependency should be explicit at the time of registration.

It can be either a singleton (one instance for all consumers) or a factory (a new instance for each consumer). We can also control whether it's synchronous or asynchronous, depending on whether we need to initialize something beforehand in an async manner. Additionally, we can specify whether it's lazy or not. Here is an example code for our use case:

lib/main.dart

```
Future<void> _setupDependencies() async {
  final getIt = GetIt.instance; // 1
  getIt.registerLazySingleton<CartRepository>(() =>
      InMemoryCartRepository()); // 2
  getIt.registerSingletonAsync<ProductRepository>(() async { // 3
    final hiveService = HiveService();
    final apiService = ApiService();
    await hiveService.initializeHive();
    return  AppProductRepository(
      remoteDataSource: NetworkProductRepository(apiService),
      localDataSource: LocalProductRepository(
        hiveService.getProductBox(),
      ),
    );
  });
  await getIt.allReady(); // 4
}

Future<void> main() async {
  await _setupDependencies(); // 5
  runApp(MaterialApp(...));
}
```

Let's review what we have done:

1. To make it more convenient, we assigned GetIt.instance to a local variable. This is the singleton that will be used for registering and fetching dependencies, essentially the service locator itself.

2. We registered CartRepository as a lazy singleton, which explicitly makes it a singleton and only creates the dependency when accessed for the first time.

3. The instantiation of ProductRepository is more complex as it requires the instantiation of various services and their initialization, which is done asynchronously. For this reason, we use the special registerSingletonAsync method, which provides a callback to perform all necessary initializations.

4. Since we have async initializations and want to ensure that all dependencies are registered before using them, we use the `await getIt.allReady();` method to wait for the initializations. If desired, this can be used in `FutureBuilder` to display a loading indicator if the setup takes too long. Ideally, we would try to defer the initialization to a later time instead of using this approach.

5. Now, we can call the `_setupDependencies` method before running our app. However, this is not a strict requirement, especially in large applications where there may be many features that won't be used. In such cases, it wouldn't make sense to initialize the entire dependency tree at the `app` start. Scoping our dependencies would be necessary in such scenarios, but we will explore this aspect in more detail in *Chapter 8*.

Now that we have registered our dependencies, how can we actually use them? The most straightforward approach would be to explicitly access them from anywhere we need. For example, we can do this in `CartBloc`:

lib/cart_bloc.dart

```
class CartBloc extends Bloc<CartEvent, CartState> {
  final CartRepository _cartRepository = GetIt.instance.get();
// Code related to cart management from the previous chapters
}
```

We use the global registry `GetIt.instance` to fetch the required dependency using the `get` method. However, this approach can result in an error if the dependency is not registered at runtime. Therefore, it is important to be mindful of this potential issue. One drawback of this approach is that it tightly couples our classes to the service locator, which is a known drawback of the pattern. To mitigate this, we can combine patterns. For example, we can keep `CartBloc` using the DI pattern by injecting dependencies in the constructor. The original implementation looked like this:

lib/cart_bloc.dart

```
class CartBloc extends Bloc<CartEvent, CartState> {
  final CartRepository _cartRepository;

CartBloc({
    required CartRepository cartRepository,
  })  : _cartRepository = cartRepository,

// Code related to cart management from the previous chapters
}
```

So, `CartBloc` continues to use the DI pattern and remains agnostic to dependency management. However, in the place where we actually create an instance of `CartBloc`, which was in `BlocProvider`, we can now use `GetIt`:

```
BlocProvider<CartBloc>(
    create: (context) => CartBloc(
      GetIt.I.get(), // 1
    )..add(const Load()),
    child: const MainPage(),
  );
```

Remember, previously in the line with the comment `// 1`, we read the required dependency via `context.read`. Now, we read this dependency via `GetIt.I.get()`. In essence, we can see that `context.read` was already an implementation of the SL pattern, with `context` acting as a central registry. The main difference was how the dependencies themselves were provided. Therefore, the DI and SL patterns are not necessarily mutually exclusive and can be combined to achieve the most flexible and pragmatic approaches. Also, by using `GetIt`, you don't need access to `context`, which can be beneficial if you need to provide dependencies where `context` is not available.

Bonus tip – using injectable

There are other tools available that we haven't discussed in this chapter, such as `injectable` (`https://pub.dev/packages/injectable`). It offers a different approach and builds on top of `get_it` but with code generation. This feature provides compile-time resolution of dependencies, which can be a game changer for big and complex projects. Neither `provider`, `flutter_bloc`, nor vanilla `get_it` offers this capability.

In a nutshell, it works in the following way:

1. Annotate your classes, such as `InMemoryCartRepository`, with special injectable annotations, for example, `@Injectable(as: CartRepository)`. Annotations for factories, singletons, and lazy singletons are available.

2. Run the `injectable` code generator, so that it generates the bindings for you, by invoking the `dart run build_runner build` command in the terminal.

3. Now, when you run the code, if any of the dependencies that you're using are not provided or are misconfigured, it won't be able to generate code, hence it won't compile! This is how you get compile-time safety for dependency initialization as a bonus, which is extremely useful as it is, and the bigger your application grows, the more potential problems this approach can help you avoid.

One thing to consider, though, is that because you have to explicitly annotate your classes, you are tightly binding yourself to the `injectable` library, and in case you ever want to change it, you would have to update a lot of files. Not necessarily an issue, but definitely something to consider.

Now that the difference between the patterns is clearer, let's take a look at when to choose which tool.

Selecting the right tool for the job

While there are many more tools available for consideration, let's compare the two tools that we have seen in practice. In general, I would suggest using the `RepositoryProvider` approach when:

- You are already using `flutter_bloc` for state management. `RepositoryProvider` seamlessly integrates with the BLoC architecture and was specifically designed for this purpose, making it a natural choice.

- You prefer scoping dependencies and managing their lifecycle within specific portions of the widget tree. Since `RepositoryProvider` is a widget, it can be added and removed from any part of the widget tree, naturally scoping its life cycle to the widget flow of the specific feature.

- Also, it is straightforward to use, especially if you are already familiar with `provider` or `bloc`.

However, if you are not using `flutter_bloc` or if you want more control over your dependencies, you can consider using other tools such as `get_it`. The benefits of selecting `get_it` are as follows:

- It is independent of your choice of state management and can integrate with any approach, providing great flexibility.

- It offers fine-grained control over the type of dependencies, such as singleton or factory, as well as their scopes, lifetimes, and async initialization.

- It is not tied to the widget tree, giving you more control and decoupling. However, be aware that it can also lead to tight coupling with the `GetIt` object itself. To mitigate this, it is recommended to combine it with the use of the DI pattern and limit the exposure to the `GetIt` service locator only when necessary. This approach will make your code easier to test and more flexible and decoupled.

In conclusion, there is more than one right answer when selecting a tool, and there are many factors to consider. It is important to perform analysis, review the pros and cons, and define your own criteria, including your must-haves and trade-offs. Feel free to explore other tools such as `injectable` to see whether they align with your project's requirements.

Summary

In this chapter, we have learned the importance of following the IoC principle in order to maintain a flexible and scalable app architecture. We have seen the benefits and limitations of various pattern implementations, such as the singleton pattern, DI pattern, and SL pattern. We obtained practical experience with various tools, such as `RepositoryProvider` from `flutter_bloc` and `GetIt` from `get_it`. Moreover, we have seen how IoC patterns in combination with the repository pattern offer a truly flexible design.

In the next chapter, *Chapter 8, Ensuring Scalability and Maintainability with Layered Architecture*, we will put together everything we have learned about our app design into architecting a scalable and maintainable structure, as well as learn even more useful principles.

Get this book's PDF version and more

Scan the QR code (or go to `packtpub.com/unlock`). Search for this book by name, confirm the edition, and then follow the steps on the page.

Note: Keep your invoice handy. Purchases made directly from Packt don't require an invoice.

8

Ensuring Scalability and Maintainability with Layered Architecture

Up to this point, we have explored various design patterns and best practices in different aspects of Flutter applications architecture. We have learned how to create a performant *View* layer, discussed options and implementations for the *Business Logic* layer, and implemented various patterns for an efficient *Model* layer. Now that we have clearly defined all our moving parts, it is time to combine them into a scalable and maintainable architecture.

Establishing efficient architecture practices and conventions is crucial to the successful life cycle of a project. It is important to understand that architecture, like everything in software, is an iterative process. In this chapter, we will review prominent architectural design patterns and apply them to our Candy Store project. This will make it productive to work with and allow for future refactoring.

By the end of this chapter, you will know how to define and split your project into layers with their own responsibilities, analyze the benefits and trade-offs of architectural decisions, and understand how following established design patterns leads to implicitly following general software engineering best practices.

The main topics covered in this chapter are as follows:

- Exploring layered architecture
- Defining layers and features
- Following software design principles

Technical requirements

In order to proceed with this chapter, you will need the following:

- Code from the previous chapter, which can be found at `https://github.com/PacktPublishing/Flutter-Design-Patterns-and-Best-Practices/tree/master/CH07/final/candy_store` and at `https://github.com/PacktPublishing/Flutter-Design-Patterns-and-Best-Practices/tree/master/CH07/final_extra/candy_store`

- All of the code required for this chapter, which can be found here:

  - Start of chapter: `https://github.com/PacktPublishing/Flutter-Design-Patterns-and-Best-Practices/tree/master/CH08/initial/candy_store`

  - End of chapter: `https://github.com/PacktPublishing/Flutter-Design-Patterns-and-Best-Practices/tree/master/CH08/final_extra/candy_store`

You can review the step-by-step refactoring in the commit history of this branch.

Exploring layered architecture

Regardless of your development background, you have most likely encountered the term **architecture** before. It would be surprising if you hadn't, since as soon as you have more than a few files in your project, you need some form of logical and navigable structure.

The primary goal of good architecture is to address two questions:

- How should you effectively separate concerns?

- How should you manage the relationships between these concerns?

Concerns can include various levels of abstractions, such as databases, APIs, user interfaces, and business logic. While there are no universally right or wrong approaches to application architecture, as it depends on the specific project and team, it is still important to establish structure and rules to follow the chosen architecture. Here are some indicators of a good architecture:

- **Separation of concerns**: The separation of concerns is essential to a solid application architecture. Clearly defining boundaries and responsibilities between different components ensures that changes in one area do not affect other areas. For example, modifying the database implementation should not impact the user interface, and changing the user interface from mobile to web should not affect the data access layers, as long as no business rules are changed. Good separation of concerns allows for independent changes in different layers.

- **Predictability and consistency**: Clearly defined separation of concerns helps developers know where to place specific components when implementing new features. It eliminates ambiguity and makes it easier to understand how different components are connected. When working on existing features, developers can easily locate the relevant components without having to consult others. Predictability and consistency are crucial to maintaining the architecture and preventing chaos.

- **Scalability and maintainability**: A predictable architecture contributes to the maintainability of a project as it grows in size and complexity. To scale efficiently, the established rules should work well at scale and allow for rapid application growth without major changes to the underlying principles. This requires upfront planning. At the same time, it is important to stay pragmatic; often, at the start of the project, we don't yet have the full requirements and complete understanding of how it will evolve. That is why we should remember that software development is an iterative process. Scalable architecture that ensures painless refactoring in the future is better than over-engineering upfront.

- **Productivity and flexibility**: Software is an ongoing process, and requirements, tools, and features continuously evolve. It is important to have a flexible architecture that can accommodate these changes without sacrificing productivity. If you find yourself spending more time fighting against the rules than creating new features, there may be an issue with the architecture.

- **Testability**: To ensure safe refactoring and guarantee the stability of the application as it grows, it is crucial to have a testable architecture and thoroughly test it on a regular basis. We will dive into testability further in *Chapter 11*.

These principles have guided us since *Chapter 3* when we started working on the Candy Store app, and we have followed them throughout the book. With many rules and patterns already in place, let's now see how to bring it all together.

Introducing multitier architecture layers

One way to separate concerns is by dividing the application into different layers. We have already touched upon this idea when discussing state management and MVVM-like patterns, where we had entities such as `View`, `Model`, and `ViewModel`. This separation of the application into layers is commonly referred to as **multitier** (or **n-tier**) architecture. The aim of this architecture is to assign specific responsibilities to each tier. While in some contexts, "tiers" may refer to physical divisions and "layers" to conceptual divisions, in this chapter, we will use the terms interchangeably, both referring to conceptual divisions. The question then becomes: how do we define these responsibilities?

In classical multitier architecture, the logic can be divided into the following layers.

Presentation layer

The **presentation** layer, also known as the view layer or UI layer, is responsible for presenting the interface to the user and handling user input.

Domain layer

The **domain** layer describes the business rules in an abstract format. It contains models that are agnostic to data sources and interfaces for repositories and services. The domain focuses on the essence of the business and what entities exist, without concerning itself with how the data is retrieved or stored.

Data layer

The **data** layer, sometimes referred to as the **infrastructure** or **persistence** layer, implements the domain layer. It contains the actual implementations of repositories and services, as well as the API and database calls.

In some cases, there may be an additional layer called the **application** or **service** layer, which acts as a bridge between the domain and presentation layers. This layer may include blocs or view models, as well as additional entities or services.

Multitier architecture is a fundamental principle of dividing application responsibilities into layers. It can be very abstract, and other architectures such as clean architecture have derived from it, introducing stricter rules and additional actors such as entities and use cases. The number of layers in an architecture can vary depending on the specific needs of the application. For example, the domain layer may be merged with the data layer in certain cases.

In the following sections, we will discuss how to split the existing code of the Candy Store app into these layers and how data flows between them, providing a clearer understanding of the architecture.

Implementing multitier architecture in the Candy Store app

The exciting part about our current task is that we already have all of the code. Now, what is left for us is to map the existing code to the respective layers. Let's do this using the example of the cart feature. In this part, we will split the files into their respective responsibilities, and in the next part, we will explore various practical implementations.

Let's start with the presentation layer. While in the original description of multitier layers, we discussed the application layer as a separate entity, in our application, it will always consist only of the bloc. Therefore, instead of making it a separate layer, we will merge it into the presentation layer. So, our presentation layer could look like this:

```
--presentation
---bloc
----cart_bloc.dart
----cart_event.dart
----cart_state.dart
---view
----cart_page.dart
---widget
----cart_button.dart
----cart_list_item.dart
```

Figure 8.1 – File structure in the presentation layer

Let's take a closer look at what we did:

1. Under the `presentation` folder, we have created three more folders: `bloc`, `view`, and `widget`.

2. The `bloc` folder is self-explanatory, as it contains only classes related to the bloc itself.

3. We have also split the UI into two categories: `view` and `widget`. We did this to easily distinguish between different levels of the UI. The `widget` folder contains small, reusable widgets that are always a part of a bigger UI. The `view` folder stores self-sustaining widgets that represent self-contained widgets, such as dialogs, full-screen pages, and so on.

Now, let's create our `domain` and `data` layers:

```
--presentation

--domain
---model
----cart_info.dart
----cart_list_item.dart
---repository
----cart_repository.dart
---service
----cart_service.dart

--data
---repository
----in_memory_cart_repository.dart
---service
----cart_api_service.dart
---model
----cart_request_body.dart
----cart_response_body.dart
```

Figure 8.2 – File structure in the domain and data layers

Let's review what we have done:

1. We have introduced a domain layer, which includes the model, repository, and service folders. In the model folder, we store agnostic data objects that contain general information about the domain models. In the repository layer, we only store the interface, which is the cart_repository.dart. The same logic applies to the service layer. Thus, the domain layer is responsible for defining domain-level logic, but not its implementation. For implementation, we have the data layer.

2. The data layer has the same folder structure as the domain layer, but it stores the actual implementations defined in the domain layer. Notice that in the model folder, we have models related to hypothetical REST API requests.

This layer structure allows us to clearly separate concerns, with each layer being responsible only for its own part of the application. It enables us to easily swap out the entire data layer for a different implementation without impacting the domain or presentation layers. This separation of concerns makes it efficient to work on different parts of the application.

Now that we understand the structure, let's review the data flow between the layers:

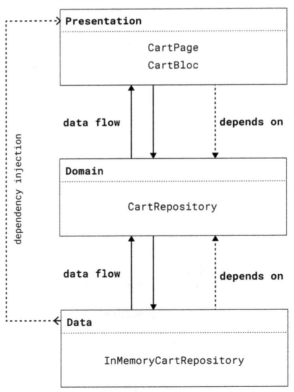

Figure 8.3 – Data flow between the presentation, domain, and data layers

Here are some important points to understand and follow:

- The domain layer is the most agnostic layer. It focuses solely on business rules and is independent of other layers. It has no dependencies or access to code in other layers.

- The presentation layer only interacts with the domain layer. It uses domain layer models and repositories (interfaces).

- The data layer also only interacts with the domain layer. It implements domain layer interfaces and maps to its models.

- The data layer is connected to the presentation layer through dependency injection mechanisms, independent of the presentation layer and is only established during **bootstrapping**. Bootstrapping refers to the initialization and configuration of our Flutter app at the moment of launching it, specifically the `runApp` method called in the `main` method. During this bootstrapping phase, dependency injection is used to provide concrete implementations (from the data layer) of interfaces (domain layer) to the presentation layer.

As you can see, the implemented patterns blend well into this architecture. This architecture is not new and is widely used in mobile and other applications on various platforms, both native and cross-platform.

Now that we have seen the conceptual separation of concerns into layers, let's explore a few different implementations and review their pros and cons.

Defining layers and features

We have introduced the concept of three layers to our architecture: data, domain, and presentation. We have seen a possible distribution of our existing files related to the cart feature, but that's just one isolated feature. Now we will review how to distribute our files when there are many features, since several approaches exist. Let's start with the simpler approach.

Implementing layer-first architecture

The first approach is known as the **layer-first** approach, also referred to as the **vertical** approach. The idea is to focus on the architecture layers first and then distribute the features inside those layers. Before we dive into the practical implementation, let's discuss the domain.

The domain layer is the most crucial layer in defining the application architecture. It remains independent of the UI (presentation) and data implementation details. Instead, it defines the model based on your business concepts, essentially a mental representation of them. This layer can represent tangible entities such as a "cart," or more abstract concepts such as a "session." Starting the architecture planning from this layer is a good idea, as implementation details can change more rapidly than the domain itself. If the domain changes, everything changes. In our example, we will define two domains: `cart` and `product`. Now, let's explore how to implement it in a layer-first manner.

First of all, we define the layers themselves:

```
--presentation

--data

--domain
```

Then, inside of *every* layer, we define our features. It could look like this:

```
--presentation

---cart

---product

--data

---cart

---product

--domain

---cart

---product
```

As you can see, we have defined our layers once, and then implemented our features within each layer. The final file structure of our current application would look like this:

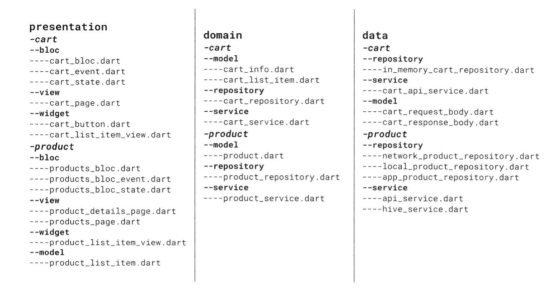

```
presentation              domain                    data
-cart                     -cart                     -cart
--bloc                    --model                   --repository
----cart_bloc.dart        ----cart_info.dart        ----in_memory_cart_repository.dart
----cart_event.dart       ----cart_list_item.dart   --service
----cart_state.dart       --repository              ----cart_api_service.dart
--view                    ----cart_repository.dart  --model
----cart_page.dart        --service                 ----cart_request_body.dart
--widget                  ----cart_service.dart     ----cart_response_body.dart
----cart_button.dart      -product                  -product
----cart_list_item_view.dart  --model               --repository
-product                  ----product.dart          ----network_product_repository.dart
--bloc                    --repository              ----local_product_repository.dart
----products_bloc.dart    ----product_repository.dart  ----app_product_repository.dart
----products_bloc_event.dart  --service             --service
----products_bloc_state.dart  ----product_service.dart  ----api_service.dart
--view                                              ----hive_service.dart
----product_details_page.dart
----products_page.dart
--widget
----product_list_item_view.dart
--model
----product_list_item.dart
```

Figure 8.4 – File structure in the layer-first architecture

This approach has both benefits and limitations. By making our layers separate packages, such as in a monorepo (see the *Exploring file structure organization in Flutter* section for more details), we can easily ensure that layer boundaries are always respected. For example, the domain layer would have its own pubspec.yaml file and would not have dependencies on data and presentation, making it impossible to accidentally import anything.

It is also intuitive to organize files according to their features, and not every feature necessarily has every layer. For example, things such as "Session" are usually not a concern of the UI and do not have a visual representation.

However, this architecture may not scale well for very large projects with numerous features and multiple developers. Working on separate features in a shared code base without interfering with someone else's domain can be challenging and often leads to code conflicts and blurred responsibilities. This concern becomes more prominent with the higher the number of people involved, rather than the number of features. Nevertheless, this simple architecture has proven to be effective in small teams.

With that being said, let's explore a more scalable and nuanced approach to layering – the feature-first approach.

Implementing feature-first architecture

The **feature-first** approach, also known as the **horizontal** approach, tackles this problem from a different angle. It prioritizes features (or domains) and then creates layers within these features. Here's an example of how this could look:

```
cart                                    product
-presentation                           -presentation
--bloc                                  --bloc
----cart_bloc.dart                      ----products_bloc.dart
----cart_event.dart                     ----products_bloc_event.dart
----cart_state.dart                     ----products_bloc_state.dart
--view                                  --view
----cart_page.dart                      ----product_details_page.dart
--widget                                ----products_page.dart
----cart_button.dart                    --widget
----cart_list_item_view.dart            ----product_list_item_view.dart
-data                                   --model
--repository                            ----product_list_item.dart
----in_memory_cart_repository.dart      -data
--service                               --repository
----cart_api_service.dart               ----network_product_repository.dart
--model                                 ----local_product_repository.dart
----cart_request_body.dart              ----app_product_repository.dart
----cart_response_body.dart             --service
-domain                                 ----api_service.dart
--model                                 ----hive_service.dart
----cart_info.dart                      -domain
----cart_list_item.dart                 --model
--repository                            ----product.dart
----cart_repository.dart                --repository
--service                               ----product_repository.dart
----cart_service.dart                   --service
                                        ----product_service.dart
```

Figure 8.5 – File structure in the feature-first architecture

As you can see, this follows the same structure we discussed earlier when talking about layers. With this approach, the features are self-contained, isolated, and can be worked on independently. If they are organized as packages, as a monorepo, instead of just folders within a single package, each feature can have its own lint rules, dependencies, and so on, making it easier to maintain them individually. However, problems arise when features start to intermingle. It is impossible to completely isolate every feature from each other – for example, certain aspects of the cart. While the cart page and checkout flows are isolated, the cart itself can be accessed from the products page and from the main page. Crossing these boundaries can lead to tight coupling, and although the idea of features being completely independent and capable of "deleting a feature package without interrupting the app flow" may sound good in theory, it is rarely the case in practice. Ultimately, you need to be pragmatic and address specific problems as they arise. Sometimes, it's acceptable for one feature to bleed into another, while other times, it may be necessary to extract it into a "common" module that is shared among features. This allows legitimate access from various features but also increases the responsibility for maintaining and managing this shared feature.

Now, let's see how we can physically organize the layers in our Flutter app.

Exploring file structure organization in Flutter

When discussing file structure, we need to consider two aspects: defining the boundaries between layers and physically distinguishing those layers. To understand this, we need to differentiate between two concepts: directories (or folders) and packages.

In a Flutter project, directories are basic organizational units used to organize code, assets, and other resources in a hierarchical structure. This structure helps manage the project more efficiently. Within the main lib directory, you can create subdirectories such as cart, bloc, view, widget, and so on. These subdirectories help segregate different parts of your application logically. Code from different files in various directories can be easily accessed using Dart imports without any additional configuration.

Here's an example of directory structure:

```
lib/
|--- main.dart
|--- cart/
|--- bloc/
|--- view/
|--- widget/
```

On a higher level, alongside your lib folder, you will find the pubspec.yaml file. This file is crucial as it manages your overall Flutter project by controlling dependencies, versioning, asset access, and more.

Here's the structure on the example of our Candy Store app:

```
candy_store/
|--- android/
|--- ios/
|--- lib/
|--- test/
|--- pubspec.yaml
```

A more advanced approach is to separate layers of your application, such as presentation, data, and domain, into distinct **packages**. In Flutter, a package is a directory that contains its own pubspec.yaml file. By introducing this file, you are forced to explicitly declare dependencies between these packages in their respective pubspec.yaml files. This method helps maintain clear boundaries between different layers of your application.

Here's how it could look in our app:

```
candy_store/
|--- presentation/
|    |--- lib/
|    |--- pubspec.yaml
 ---
|--- data/
|    |--- lib/
|    |--- pubspec.yaml
 ---
|--- domain/
|    |--- lib/
|    |--- pubspec.yaml
 ---
|--- pubspec.yaml (root level)
```

> **Good to know**
>
> Storing multiple packages in a single repository is known as the **monorepo** approach. This is particularly beneficial to reusable packages or when multiple teams are collaborating on a single project. It allows different packages to be developed simultaneously while being part of one product, which is especially useful if following a feature-first approach to app architecture. A popular tool for managing a monorepo in Flutter is **Melos** (https://pub.dev/packages/melos).

Before we move on to learning about scoping, it is worth noting that in the "ideal" architecture, there is a clear distinction between domain models, data models, and presentation models. However, this often results in duplicated code and unnecessary mapping in real-world scenarios. There is no need to create models solely for the purpose of strictly adhering to architectural principles. For instance, ideally, we would have a separate `CartListItem` model in the `presentation` layer, called `DisplayCartListItem`, which could contain different information than the `domain`-level `CartListItem`. But in our scenario, they are the same, so we can simply use the same model as is. The same applies to other mappings, such as serialization to and from JSON – while, in theory, the API can change from returning responses in JSON format to returning them in Protobufs, in reality, it's rarely the case, so we could merge this aspect of the data layer with the domain layer. Always remember to stay pragmatic and keep things simple while you can, since adding complexity often introduces more problems than it solves.

Scoping dependencies to a feature life cycle

By following the feature-first approach, we can clearly identify the entry point for a feature. If a feature is self-contained and carefully scoped, it is possible to dynamically scope its dependencies. For example, up to this point, we have defined all of our dependencies in the main.dart file using either RepositoryProvider or GetIt, which makes their scope at the application level. This means that all dependencies are created when the application is launched and they persist until the app is closed. This approach works fine for features such as the cart and products, as their life cycles are tied to the app. However, as our app grows, our features become more conditional.

For instance, the user may add items to the cart at any time, but it doesn't necessarily mean they will proceed to the checkout. Therefore, it would be impractical to create all the dependencies related to checkout during app startup, especially if their initialization takes time and slows down the app's launch. Additionally, it would clutter the main method with an increasing number of dependencies as the app expands.

Instead, we can create a new feature called checkout and initialize its dependencies only when it is first launched. For example, using RepositoryProvider from flutter_bloc, we can wrap our CheckoutPage in a MultiRepositoryProvider and only create the CheckoutRepository there. Furthermore, the CheckoutRepository will be removed once we navigate away from the CheckoutPage, effectively binding its life cycle to that of the feature. You can also redefine dependencies if needed in this context. To make the code boundaries even more well-defined, we can introduce a CheckoutFlow. This way, we will know that a Flow notion initializes a set of scoped dependencies. So, in order to launch the checkout flow, we can use the following method where we need it:

lib/cart/presentation/view/cart_page.dart

```
void _initCheckout() {
    Navigator.of(context).push(
      MaterialPageRoute(
        builder: (_) => const CheckoutFlow(),
      ),
    );
}
```

Here, we simply launch the CheckoutFlow via Navigator, and the CheckoutFlow itself is a simple widget that wraps the CheckoutPage in the MultirepositoryProvider:

lib/checkout/presentation/view/checkout_flow.dart

```
class CheckoutFlow extends StatefulWidget {
  const CheckoutFlow({super.key});
```

```
    @override
    State<CheckoutFlow> createState() => _CheckoutFlowState();
}

class _CheckoutFlowState extends State<CheckoutFlow> {
    @override
    Widget build(BuildContext context) {
      return MultiRepositoryProvider(providers: [
        RepositoryProvider<CheckoutRepository>(
          create: (context) => StubCheckoutRepository(),
        ),
      ], child: CheckoutPage.withBloc());
    }
}
```

As you can see, we provide the `CheckoutRepository` implementation via the `StubCheckoutRepository` class. We can then change it only in here, and the whole dependency scope is bound to the life cycle of the `CheckoutFlow` widget.

With `GetIt`, it is also possible to scope the life cycle of dependencies. This can be achieved by using the `pushNewScope` and `popScope` methods, similar to how you would navigate with a `Navigator`. Let's look at the code:

lib/cart/presentation/view/cart_page.dart

```
Future<void> _initCheckout() async {
    GetIt.instance.pushNewScope( // #1
      scopeName: 'checkout',
      init: (scope) {
        scope.registerSingleton<CheckoutRepository>(
          StubCheckoutRepository(),
        );
      },
    );
    await Navigator.of(context).push( // #2
      MaterialPageRoute(
        builder: (_) => const CheckoutFlow(),
      ),
    );
    GetIt.instance.popScope(); // #3
}
```

We have done a couple of things in this code snippet:

1. Via the `GetIt.instance` method `pushNewScope`, we have initialized and pushed a new tree of dependencies, which currently consists only of `CheckoutRepository`.

2. Then, we pushed the already familiar `CheckoutFlow` onto the navigation stack.

3. We wait until the `CheckoutFlow` has been popped off the navigation stack, and after that, we also pop the `GetIt` scope, which will remove the topmost dependency scope from the dependency stack.

As you can see, you can use various tools to achieve the same goals. There is often more than one correct solution to a problem. Let's review another such problem, which is what to do if we have several data sources that need to interact.

How to connect multiple data sources

There may be situations where our screens rely on more than one data source, as we have seen with our "Product" implementation. If we have products saved in the cache, we show the cached value to the user. If not, we fetch them from the API. This process can become complex, involving considerations such as internet connection, cache invalidation, and the type of internet connection (Wi-Fi or mobile data). There are several approaches to handling this complexity:

- As we saw in *Chapter 6*, we can create an "umbrella" repository called `AppProductRepository` that encapsulates our specialized data sources (API and Hive), providing a consistent interface to the consumer. This is a viable approach with no issues.

- However, you may also want to differentiate between repositories with single responsibilities and introduce another level of abstraction: a service. The service encapsulates multiple repositories and their relationships, and the consumer interacts with them exclusively through the service.

- You can combine these approaches. In your blocs or view models, use a repository when it is sufficient, or a service when needed.

- If you want to take it further, instead of creating a single service that handles multiple functionalities, you can split it into use cases based on single functions. For example, you could have a `LogoutUseCase` that accepts all the repositories with data as constructor parameters and, when executed, clears all these repositories. This UseCase can then be used in blocs or view models, as defining this single function in an abstract class of `LogoutRepository` might be excessive.

- Alternatively, if you want to keep it minimal or have highly specific logic for a single bloc or cubit, you can encapsulate it within the bloc or cubit itself, without delegating anything to the repository, service, or use case.

This overview of possible solutions to the same problem highlights the wide range of challenges that can arise. There is rarely a one-size-fits-all solution, and while it is important to establish conventions and adhere to them as much as possible, it is also important to be pragmatic. Remember that patterns and conventions should serve you, and if they make things more difficult, it is reasonable to review them and use what makes the most sense. Be strict yet flexible and pragmatic. The architecture is meant to serve you, not the other way around.

Before concluding this chapter, let's review another important concept: software design principles.

Following software design principles

Up to this point, we have reviewed various patterns at different levels, ranging from designing a state management approach to a full-blown app architecture. One observation you may have made is that many of these patterns are not specific to Flutter. Architectures such as MVVM and MVI are common in various software domains, as are concepts such as dependency injection and n-tier architecture. However, we have adapted these patterns to suit the specific intricacies of Flutter.

In the next chapters, we will dive deeper into the specifics of Flutter. Before we do that, however, I would like to discuss one last topic: how to apply general software design principles to Flutter.

Deciphering the acronyms of SOLID, DRY, KISS, and YAGNI

If you have been in software development for some time, chances are you have heard about some or all of these acronyms when discussing code quality and architecture decisions. Programmers seem to be fond of acronyms and catchy names for design guidelines, such as the "Hollywood Principle" or the "Tell, Don't Ask Principle." Let's review how we have been following these principles all along.

Simplifying code with DRY, KISS, and YAGNI

The principle *DRY* stands for *don't repeat yourself*, *KISS* stands for *keep it simple, stupid*, and *YAGNI* stands for *you ain't gonna need it*. They are abstract and can have various interpretations, but in general, they serve as reminders against over-engineering. Discussions about architecture often involve concerns about over-engineering, which are completely valid. Premature optimization can result in throwaway code and wasted time. For example, when deciding not to duplicate the `CartListItem` into a `DisplayCartListItem`, we have followed all three principles! By avoiding the creation of a duplicate data model for the sake of abstractions, we have normalized the data and followed DRY. We also created a class that we don't need yet, following YAGNI. Overall, we kept it simple, adhering to KISS. As you can see, these are not strict rules to memorize, but rather qualities that naturally emerge when applying a pragmatic approach to decision-making.

Following the SOLID principles

SOLID is an acronym that stands for **single responsibility, open-closed, Liskov substitution, interface segregation, dependency inversion** software design principles. The SOLID principles are often both praised and criticized (as outdated and irrelevant). However, I believe that keeping these principles in mind has helped me make better choices in architecture and code design. It's important to note that these principles are guidelines rather than strict rules that must be followed blindly. Now, let's review how we have applied these principles in our Candy Store app.

The single responsibility principle

This principle states that a class should have only one reason to change. For example, we have the `NetworkProductRepository` and the `LocalProductRepository`. The `NetworkProductRepository` only needs to change if there are any modifications in the API calls, while the `LocalProductRepository` operates independently. In our generic `AppProductRepository`, changes are only required if the caching algorithm is modified, not if there are changes in the API or local storage preferences.

The open-closed principle

This principle states that software entities should be open for extension but closed for modification. In our case, this principle is relevant to interfaces. We work with abstract classes (interfaces) that can be extended and allow us to create and swap different implementations. For example, we have the abstract `CartRepository` and during development, we use the `InMemoryCartRepository` implementation. However, this implementation can be easily substituted with the `NetworkCartRepository` without modifying the consumer code.

The Liskov substitution principle

This principle states that an object or class can be replaced by a sub-object or subclass without breaking the program. In our case, we adhere to this principle with the `ProductRepository`. All of our blocs use the `ProductRepository` interface, allowing us to substitute the implementation with a generic `AppProductRepository` that includes a caching algorithm. We can also use subclasses such as `LocalProductRepository`, which fetches products only from local storage without any impact on the program's behavior.

The interface segregation principle

This principle states that code should not be forced to depend on methods it does not use. In other words, interfaces should be small and specialized. In our feature-first architecture, we have separate `CartRepository` and `ProductRepository` instead of a single large `StoreRepository`. Each repository operates with a small set of methods that are relevant only to its domain.

The dependency inversion principle

This principle has two parts:

- High-level modules should not depend on low-level modules. Both should depend on abstractions (interfaces).

- Abstractions should not depend on details. Details (concrete implementations) should depend on abstractions.

We follow this principle by introducing the data and domain layers into our architecture. The domain module (high-level module) does not depend on the data module (low-level module) and does not import anything from it. Our concrete implementations, such as the blocs, only depend on abstractions such as `ProductRepository` and `CartRepository`.

Just for fun!

We have discussed some of the most popular acronyms and catchy principles, but there are countless others. For instance, there is the *WET* principle, which is the opposite of the DRY principle, and there are also acronyms such as *AHA*, among many others. Some of these acronyms are meant as jokes, while others are more serious. Just for fun, try looking them up and see how many you can find. Who knows, maybe some of them will even be useful!

As you can see, even without prior knowledge of these principles, we have gradually followed patterns and created an architecture that implicitly adheres to these widely adopted guidelines.

Summary

In this chapter, we have learned the importance of establishing and maintaining a scalable app architecture. We have learned how to scope and manage the responsibilities of the application's moving parts by adopting the multitier architecture and its presentation, data, and domain layers. We have seen various approaches and reviewed their benefits and trade-offs, which equipped us with the knowledge to make the correct choice for ourselves. We have also seen that there is not just one correct solution, but rather multiple approaches that can be taken. It is important to apply critical thinking and try different approaches in practice to determine what works best for you.

Additionally, it is important to follow good design practices. We reviewed how, by following best practices and design patterns, we have implicitly followed the widely adopted good software design principles, such as DRY, KISS, YAGNI, and SOLID.

In *Chapter 9*, *Mastering Concurrent Programming in Dart*, we will dive into the world of Flutter concurrency and learn how to work with asynchronous tools of the Dart language, as well as how to handle parallelism via Flutter isolates.

9

Mastering Concurrent Programming in Dart

In the previous chapters, we explored different design patterns and architecture best practices in Flutter. Now, it's time to explore another important topic: working with **concurrency** in Dart.

Flutter is known for catchy taglines such as "Everything's a Widget" and "Dart is single-threaded." We have already debunked the former and learned that not everything is a widget as there are also elements and render objects. In this chapter, we will investigate the truth behind the saying "Dart is single-threaded." We will explore the asynchronous world of Flutter and Dart, understand the peculiarities of their APIs, and learn how to work with them correctly and efficiently.

Concurrency-related concepts are often complex, leading to confusion, misuse, and frustration in programming. However, they are crucial because modern applications and programming languages require concurrency support to implement a wide range of features and behaviors.

But don't worry! In this chapter, we will gradually learn about the concurrency model of the Dart language and Flutter applications, and discover how to work with it efficiently.

The following main topics will be covered in this chapter:

- Dart is single-threaded. Or is it?
- Working with Future APIs in Dart
- Embracing parallelism with isolates

Technical requirements

To proceed with this chapter, you will need the following:

- The code from the previous chapter, which can be found here: `https://github.com/PacktPublishing/Flutter-Design-Patterns-and-Best-Practices/tree/master/CH08/final/candy_store`

- The code required for this chapter, which can be found at the following links:

  - Start of this chapter: `https://github.com/PacktPublishing/Flutter-Design-Patterns-and-Best-Practices/tree/master/CH09/initial/candy_store`

  - End of this chapter: `https://github.com/PacktPublishing/Flutter-Design-Patterns-and-Best-Practices/tree/master/CH09/initial/candy_store`.

You can review the step-by-step refactoring in the commit history of this chapter's branch.

Dart is single-threaded. Or is it?

There's a lot to unpack in this three-word sentence. At first glance, it might seem like an easy question with a simple "yes" or "no" answer. However, taking such a simplistic approach will not help us achieve our goal. Answering this question requires that we discuss concepts such as concurrency, synchronicity, threads, parallelism, and blocking. While it may seem overwhelming if you haven't encountered these concepts before, we will explore them gradually and logically, building our knowledge and experience from the ground up.

Understanding synchronous, concurrent, and parallel operations

There's nobetter way to understand how the code works than writing, executing, and examining the results. Let's start by executing the following code:

lib/examples/example1.dart

```
void main() {
  print("1");
  print("2");
  print("3");
}
```

Even without running the code (and then by actually doing it and verifying it), we can guess that the output will be as follows:

```
1
2
3
```

The output we can see was produced synchronously. We call the functions one by one, one after another, so we see the expected result. The function calls are processed synchronously. We can show it schematically like this:

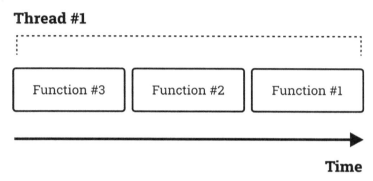

Figure 9.1 – A schematic diagram of how the events are processed synchronously on a single thread

This is an easy scenario. But in modern programming, we are used to handling more complex scenarios, and for more complicated problems, we need to provide more efficient solutions. Let's say we want to read a file from a disc. This operation can take time – we need to query the filesystem, open it, read data from it, and then close it. If we give all our resources to this task, our UI will be unresponsive until it's done, which can take seconds or even longer. And we can be sure that nobody wants that. One way to approach this problem is to split the task into many smaller tasks, and in between them handle other requests, such as screen rendering. Schematically, it could look like this:

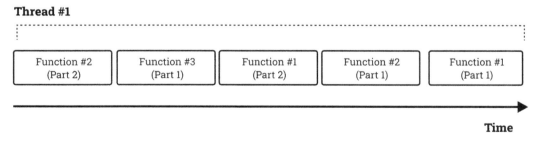

Figure 9.2 – A schematic diagram of how the events are processed asynchronously on a single thread

As you can see, in this case, the events are still processed *sequentially* and *on the same thread*. It's just that the bigger event itself is split into many more, and new events can start before the previous ones are finished. So, in this approach, the events are processed asynchronously or concurrently. Concurrency is often confused with parallelism as developers assume that to process events asynchronously, they must be processed in parallel. As we have seen, this is not the case. Keeping this in mind as we dive into the Dart specifics of concurrency is crucial.

Now, let's imagine the following scenario: your app needs to show a loading animation, read a file from disk, and perform a network operation – preferably all at once because these actions don't depend on the results of one another. In other words, they can be processed in parallel. This approach would require launching operations on different threads, leading us into multi-threading territory. Multi-threading is a potent tool as it allows several tasks to be executed simultaneously, but it also has its peculiarities and risks if misused. Schematically, this process would look like this:

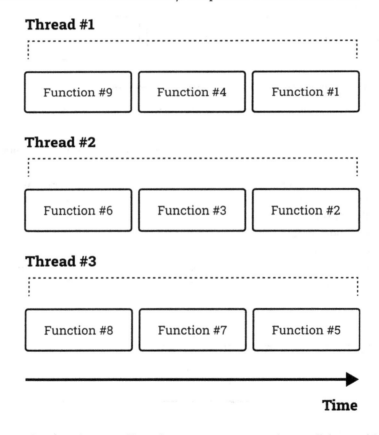

Figure 9.3 – A schematic diagram of how the events are processed in parallel on multiple threads

With that, we've discussed the fundamental concepts at an abstract level. Flutter supports these concepts. Now, let's look at how Flutter implements them. We will start with the events happening in the application synchronously – for example, "user enters value → clicks button → new value is rendered on the screen." But how does Flutter know which events to process in which order? For that, we have the event loop.

What is the event loop?

The concept of the **event loop** is not unique to Flutter. It refers to a process that takes a queue of events and processes them one by one. To execute any action, it must be added to the event processor, known as the event loop. Fortunately, this is handled automatically in Dart (and, by extension, Flutter). However, understanding the inner workings of the event loop is important to write efficient code and avoid potential bugs.

Schematically, the event loop works like this:

Figure 9.4 – A schematic diagram of how the event loop works

It is important to note that the events in the event queue are processed in the order they are submitted, following the **first in, first out** (**FIFO**) principle of computer science data structures. In other words, they are processed sequentially.

Now, this is where things start to get interesting. In real-world applications, it is impossible to process everything synchronously. After all, we have animations, network requests, and database accesses that often seem to happen simultaneously from the user's perspective. There must be a mechanism that allows for this; otherwise, we would end up with frozen screens and unresponsive apps... And of course, such a mechanism in Dart exists.

Understanding blocking operations in Flutter

In *Chapter 1*, we discussed the build method of the widget, which needs to execute within 8 to 16ms (depending on the device refresh rate). This is because the build method can potentially be called 60 to 120 times per second. These calls can occur for various reasons, such as when the user invokes setState or when the Flutter engine itself triggers them. Now that we understand the event loop, we know that the engine can potentially post an event to build the whole visible widget tree every

8ms, for example, on an animation tick. But what happens if we fail to meet this requirement? Let's consider an example. Here is a simple application that displays an indefinitely spinning progress widget:

lib/examples/example2.dart

```dart
void main() {
  runApp(const ProgressWidget());
}

class ProgressWidget extends StatelessWidget {
  const ProgressWidget({super.key});

  @override
  Widget build(BuildContext context) {
    return const MaterialApp(
      home: Scaffold(
        body: Center(child: CircularProgressIndicator()),
      ),
    );
  }
}
```

In this scenario, everything is working fine. However, if we tweak this code by adding a couple of lines and rerun the app, we will notice that the screen completely freezes for a certain period. Here's the problematic code:

lib/examples/example3.dart

```dart
class ProgressWidget extends StatelessWidget {
  const ProgressWidget({super.key});

  @override
  Widget build(BuildContext context) {
    _block();

    return const MaterialApp(
      home: Scaffold(
        body: Center(child: CircularProgressIndicator()),
      ),
    );
  }

  void _block() {
```

```
    for (int i = 0; i < 1000000; i++) {
      print('$i');
    }
  }
}
```

Here, we have added a `for` loop with 1,000,000 iterations in our `_block` method and called it inside the `build` method. On each iteration, it prints the number before returning the actual widget. The reason the screen is frozen is that this operation takes more than 16ms, and the subsequent frames are not getting rendered (they are being dropped) because we have blocked the thread responsible for rendering. In this scenario, it is a blocking operation, meaning that until it is done, the executor is blocked from doing anything else, even if it means skipping crucial operations such as rendering.

First, this serves as a demonstrative reminder of why you should not perform any number crunching in the `build` method. Secondly, it shows that we can indeed block the main thread. Technically, everything in Dart and Flutter runs on a single thread known as the main thread. Blocking it will result in various unwanted results and behaviors.

Here is another example to demonstrate that this issue is not only present when called inside the `build` method. Let's run the same `for` loop outside of the `build` method, in a callback that's triggered when we click on the progress indicator. Here's the code:

lib/examples/example4.dart

```
class ProgressWidget extends StatelessWidget {
  const ProgressWidget({super.key});

  @override
  Widget build(BuildContext context) {
    return MaterialApp(
      home: Scaffold(
        body: Center(
          child: GestureDetector(
            onTap: _block,
            child: const CircularProgressIndicator(),
          ),
        ),
      ),
    );
  }

  void _block() {
    for (int i = 0; i < 1000000; i++) {
      print('$i');
```

```
      }
    }
  }
```

If you run it and click on the progress indicator, you will notice that it remains frozen until the loop completes, resulting in the same outcome as before. However, there are various approaches available to mitigate this issue. Let's discuss concurrency in Flutter.

Working with Future APIs in Dart

Up until this point, we have been working with synchronous APIs, meaning everything was processed in the same order it was requested, and nothing was processed before the previous event had been handled. However, we have noticed that this approach sometimes blocks the main thread due to greedy operations. Additionally, we observed that all operations were performed on a single thread, the main thread. Moving forward, we will continue to work with this main thread, but we will learn how to prevent thread blocking by utilizing asynchronous APIs.

What does the Future hold?

If you have worked with Flutter, you have likely encountered the concept of `Future`. If you are new to this concept, don't worry – we will cover it from the beginning.

The concept of the `Future` class is not unique to Dart. This notion in computer science, also known as promise, delay, or deferred (`https://en.wikipedia.org/wiki/Futures_and_promises`), describes concepts that are used to organize tasks in programs that do things concurrently. They serve as placeholders for unknown results while they are still being calculated. This allows us to decouple the value of the function from its computation so that we can implement asynchronous operations. Here are a couple of important concepts to understand about Futures:

- A Future can be uncompleted, meaning the value hasn't been computed yet, or completed. Completed is also referred to as "resolved" and can either be completed with a value or an error.

- A Future can be executed in a blocking or non-blocking way. For example, you can wait for the resolution of the Future in a blocking way, or synchronously, and block any operations until this Future is resolved. You can also continue processing without awaiting the resolution of the Future or by registering callbacks, hence executing it in a non-blocking way.

- The evaluation strategy of the Future can be either eager or lazy. Eager means that the computation of the Future starts at the moment of its creation. Lazy evaluation occurs when the first value is accessed. In Dart, Futures are evaluated eagerly.

Now that we have overviewed the abstract concept of a Future in computer science, let's see how this can be achieved and how it can be useful in Dart specifically.

To address the previous problem of blocking the UI thread, we can modify our _block function, make it return a Future, and introduce a 1ms delay before every print statement. The code will look like this:

lib/examples/example5.dart

```
class ProgressWidget extends StatelessWidget {
  const ProgressWidget({super.key});

  @override
  Widget build(BuildContext context) {
    return MaterialApp(
      home: Scaffold(
        body: Center(
          child: GestureDetector(
            onTap: _block,
            child: const CircularProgressIndicator(),
          ),
        ),
      ),
    );
  }

  Future<void> _block() async { // 1
    for (int i = 0; i < 1000000; i++) {
      await Future.delayed(const Duration(milliseconds: 1)); // 2
      print('$i');
    }
  }
}
```

If you run this code and tap the progress indicator, you will see that everything works as expected. The for loop is executed while the progress indicator is animated and not frozen. So, what did we do? We introduced a Future class into our code in two places:

1. In the signature of the _block function, we changed the return type from void to Future<void> and marked it with the async keyword.

2. Before printing the value of i, we used the await keyword and applied it to a Future. delayed constructor, passing Duration with 1ms as the delay duration.

Keep in mind that this code serves only for demonstration purposes. There are better methods for "number crunching" tasks, meaning CPU-heavy operations, which we will discuss later in this chapter, in the *Embracing parallelism with isolates* section. Another important thing to note is that if we remove the `await` keyword on line 2, our problem will return! This is because just marking the `_block` method with the `async` keyword *is not enough*. We need to *explicitly* await a `Future` result for the rest of the code to be performed *after* the resolution of the `Future` instance.

There is a lot to understand here, and we will explore the Future API in detail. But before that, let's visualize the difference to get a better understanding. Previously, without using a `Future` class, our `for` loop posted a single large event onto the event loop. The event loop was completely occupied with processing this event and couldn't attend to any other requests, resulting in frame drops. Visually, it looked like this:

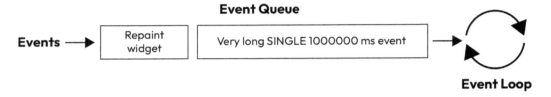

Figure 9.5 – The event loop being blocked by a single event with a duration of 1,000,000ms

However, by awaiting a `Future` class with a `zero` duration, we divided the events into 1,000,000 distinct events for the event loop. This allows the event loop to handle other requests in between. Although it appears instantaneous to us, it's important to remember that 1 second for us humans is enough time for the computer to execute thousands or even millions of operations. Visually, it now looks like this:

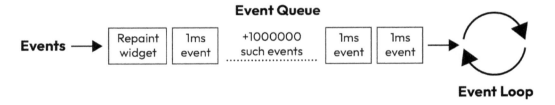

Figure 9.6 – The event loop processing 1,000,000 single events with a duration of 1ms

So, technically, if the engine requested a widget repaint due to an animation tick, for example, it could've posted that event onto the event loop in between the 1ms events and there would be no frame drop. Note that 1ms is an arbitrary value and we could change it to a smaller or a larger duration. For example, it could look like this:

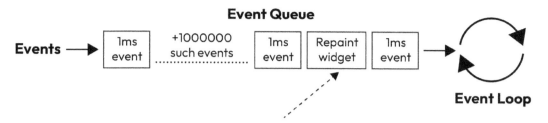

Figure 9.7 – A "repaint widget" event posted on the event loop in between 1ms events

Another critical thing to note here is that the events are still processed sequentially, one after another, and on a single thread, despite introducing the Future concept and using the async and await keywords, which imply asynchronous behavior. Let's clarify this confusion.

Understanding the concept of async

One of the most confusing things is the concept of **asynchronicity**. It is often assumed that asynchronous processing involves parallel execution with multiple threads. This is opposed to synchronous processing, which occurs sequentially on a single thread. However, as we discussed previously, this is not necessarily the case, especially in the context of Futures in Dart. The code inside a Future is still executed on the same main thread, but it allows the following to occur:

- Other code can be executed while waiting for the Future to complete

- The main thread is not blocked while waiting for an operation to finish

If this explanation has confused you even more, don't worry. Let's see some practical examples to make these points clearer. First, let's run this simple piece of code:

lib/examples/example6.dart

```dart
void main() {
  print("Before"); //1
  printWith1msDelay(); //2
  print("After"); //3
}

Future<void> printWith1msDelay() async {
  await Future.delayed(const Duration(milliseconds: 1));
  print("Print with 1ms delay");
}
```

> **Tip**
> Before running the code, try guessing the order in which the `print` statements will be displayed. Did you get it right?

Maybe you expected to see this result:

```
Before
After
```

After all, we didn't `await` the async function, `printWith1msDelay`. Let's address this idea first. Recall that we discussed eager versus lazy evaluation of the `Future` class at the beginning of this section. In Dart, Futures are evaluated eagerly, which means as soon as they're created. They don't require `await` or any special trigger functions to be executed. In general, a `Future` return type simply means that a function returns a `Future` class, similar to how it would return a `String` or `Object` class. But it gets evaluated right away. This is the first thing you should understand and remember. It's a common misconception that if you don't call `await` on the `Future` class, it won't start executing.

That being said, perhaps you were expecting to see this:

```
Before
Print with 1ms delay
After
```

Since, in the main function, we are calling these functions one after another, you already know that we don't need to `await` the `Future` instance. The thing is, when you create a `Future` instance, it is added to the event loop. As it is inherently asynchronous, it may take some time to execute. So, the `Future` instance is immediately posted on the event loop, and the callback is registered in the callback registry (callback registry is mentioned as a term here: `https://dart.dev/language/concurrency#event-loop`). However, the callback itself will only be executed during the next iteration of the event loop. So, what we actually see is this:

```
Before
After
Print with 1ms delay
```

What we have observed here is that by utilizing the Future API, we can make our code asynchronous. This means that once an asynchronous operation starts, other operations can also start and end before the initial operation finishes. The sequence of starting and ending is not sequential, but it is also not parallel. On the event loop, only one action is processed at a time. Here is a schematic representation from the perspective of the event loop:

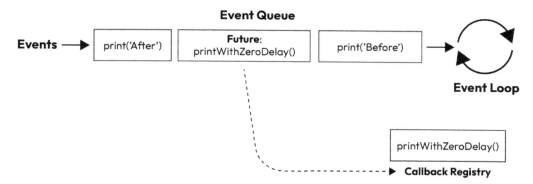

Figure 9.8 – A Future posted on the event loop and its callback registered in the callback registry

This diagram shows that when the Future was created, it was posted on the event loop synchronously. Yet the callback itself wasn't executed then – it was only registered in the callback registry. The event loop then went to process the next event in the queue – the `print('After')` event. Finally, on the next iteration, the callback itself gets triggered and removed from the callback registry:

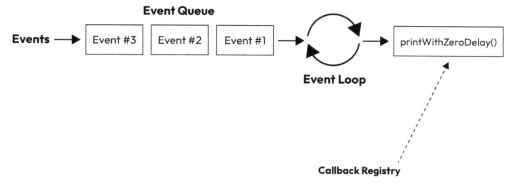

Figure 9.9 - The printWith1msDelay callback executed and popped from the callback registry

Futures in Flutter

As you work with the Flutter framework, you will see Futures being used for everything. For example, if you use `Navigator` to navigate to a page via a push function such as `Navigator.of(context).push(...)`, you will see that it returns a `Future` class. This allows you to track when the page you just pushed gets popped off the stack and any potential return values. Another example is reading from a `File` class. The `File` class has methods that return a `Future` class, such as `readAsBytes()`, which returns `Future<Uint8List>`, or `readAsBytesSync()`, which returns just `Uint8List`, not wrapped in a `Future` class and performs the read operation synchronously.

As we can see, Future is just a mechanism to execute asynchronous operations, yet on the same thread and not in parallel. But what happens when we have multiple Futures to deal with?

To chain Futures or to await, that is the question

We can modify the flow by introducing synchrony into our asynchronous code. Let's update our code so that it looks like this:

lib/examples/example7.dart

```
void main() async {
  print("Before"); // 1
  await printWith1msDelay(); // 2
  print("After"); // 3
}

Future<void> printWith1msDelay() async {
  await Future.delayed(const Duration(milliseconds: 1));
  print("Print with 1ms delay");
}
```

Upon running it, we will observe the expected result:

```
Before
Print with 1ms delay
After
```

This occurs because we have included the `await` keyword before our asynchronous `printWith1msDelay` function. By using the `await` keyword, we notify the Dart runtime that it should await the completion of this function's execution before proceeding with the code. Once the associated `Future` class has finished, the code will continue executing from this point onward. Similar behavior can be achieved by chaining our Futures in the following manner:

lib/examples/example8.dart

```
void main() {
  print("Before"); // 1
  printWith1msDelay()
      .then((_) => print("After")); // 2
}
```

We can chain multiple futures one after another, just as we would `await` them sequentially.

> **Good to know!**
>
> The event queue is not the only queue of actions that the event loop processes. After every event, the event loop also checks a queue with higher priority – the microtask queue. If this queue is not empty, all the events on the microtask queue are executed before returning to the event queue. This queue is necessary to control the execution flow and prioritize tasks. For example, it ensures that the `then` method of a Future is executed immediately after the initial Future completes. This can be achieved by posting the completion event on the microtask queue.
>
> Keep in mind that a Future can be in one of two states: uncompleted or completed. Once it is completed, it can have either a value or an error. As a developer, it is unlikely that you will need to manually post anything on the microtask queue. Even the Flutter framework rarely uses this queue, and only for critical tasks. Overloading the microtask queue can result in the event queue being starved, which can lead to dropped frames and other performance issues.

After considering these two options, a logical question arises: which one is the preferred approach to use? There is no definitive answer as it can depend on the specific use case, error handling requirements, as well as personal and team preferences. However, as a general rule of thumb, it is generally better to utilize the `async`/`await` syntax and resort to Future chaining only when it offers functional advantages, such as more precise error handling. This is because code written in a more linear style, such as with `async`/`await`, tends to be visually more straightforward to read and understand. Since we spend more time reading code than writing it, readability is an important thing to consider when selecting an approach.

Speaking of error handling, Future chaining not only allows us to chain futures using `then` but also enables us to add error listeners on specific futures using `catchError`. For instance, our code may appear as follows:

lib/examples/example9.dart

```dart
void main() async {
  print("Before");

  printWith1msDelay()
      .then(
        (_) => printWith1SecondDelay().catchError( (ex, st) {
        print("Handle error thrown by printWith1SecondDelay()");
      },
    ),
  )
      .catchError((ex, st) {
    print("Handle error thrown by printWith1msDelay()");
```

```
    });
  }
```

While the code becomes increasingly harder to read with more chaining involved, it does give the benefit of addressing errors of specific Futures in their error handlers. With the `async/await` syntax, we would need to catch each exception on its own in the topmost error handler via `try/catch`, something like this:

lib/examples/example10.dart

```dart
void main() async {
  print("Before");
  try {
    await printWith1msDelay();
    await printWith1SecondDelay();
  } on ImportantException catch (ex, st) {
    print("An important exception was thrown");
  } on UnimportantException catch (ex, st) {
    print("An unimportant exception was thrown");
  }
}

Future<void> printWith1msDelay() async {
  await Future.delayed(const Duration(milliseconds: 1));
  print("Print with 1 ms delay");
  throw UnimportantException();
}

Future<void> printWith1SecondDelay() async {
  await Future.delayed(const Duration(seconds: 1));
  print("Print with 1 second delay");
  throw ImportantException();
}

class ImportantException implements Exception {}
class UnimportantException implements Exception {}
```

Here, we use multiple `catch` clauses. This allows us to catch specific exceptions, but we don't always know about their origins. However, it also doesn't always matter – for example, we will want to show the user a message when they are logged out due to an error: it doesn't matter if the subtype was `InvalidAuthException` or `ExpiredAuthException`, just that it was an instance of a more abstract `AuthException`. On the other hand, if the user were trying to purchase something and an error occurred, we would want to tell them the cause and how they could fix it, so a more granular error-handling approach would be useful in this case. So, as usual, the best practice is to be pragmatic. Now that we know the principles of working with Futures and efficient error handling, let's take a look at strategies for working with multiple independent Futures.

Handling independent Futures efficiently

The Future class provides powerful APIs for handling asynchronous operations. Sometimes, there are scenarios where one Future depends on the result of another. In such cases, we need to await the completion of the first Future before moving on to the next. For example, consider the following scenario:

lib/examples/example11.dart

```dart
void main() async {
  final data = await getImportantData();
  await printImportantData(data);
}

Future<String> getImportantData() async {
  await Future.delayed(Duration(seconds: 1));
  return 'Important data';
}

Future<void> printImportantData(String data) async {
  await Future.delayed(Duration(seconds: 1));
  print(data);
}
```

In this scenario, we cannot call printImportantData without passing the data parameter. Therefore, we need to await the completion of getImportantData before moving on.

However, let's consider a different scenario:

lib/examples/example12.dart

```dart
void main() async {
  final sw = Stopwatch()..start(); // 1
  await doSomeImportantWork();
  await doSomeImportantWork();
  await doSomeImportantWork();
  sw.stop();
  print('Total time: ${sw.elapsedMilliseconds}ms'); // 2
```

```
  }

Future<void> doSomeImportantWork() async {
  await Future.delayed(Duration(seconds: 3));
  print('Important work done');
}
```

In this scenario, we are awaiting three method calls to doSomeImportantWork, which in our imaginary scenario perform some very important work that takes 3 seconds each time. We use Stopwatch on line 1 to measure the total time of the operation. When we print the result (line 2), we will see a total time of around 9 seconds.

However, these method calls do not depend on each other. Therefore, there is a more efficient way to handle this using Future.wait. Let's see it in practice:

lib/examples/example13.dart

```
void main() async {
  final sw = Stopwatch()..start();
  await Future.wait([
    doSomeImportantWork(),
    doSomeImportantWork(),
    doSomeImportantWork(),
  ]);
  sw.stop();
  print('Total time: ${sw.elapsedMilliseconds}ms');
}
```

The Future.wait constructor receives an Iterable class of Futures and initiates their execution simultaneously. However, it is important to note that Futures are not executed in parallel. Instead, for each doSomeImportantWork, the 3-second timer callback is registered one by one in the callback registry. Since they are all registered almost at the same time (from a human perspective), they will all be complete in a little over 3 seconds.

What Future.wait does is kickstart all of the Futures one by one without waiting for the previous one to complete. It only completes once all of the Futures that have been passed to it are complete. If any of the Futures throws an error, that error will be the result. If multiple Futures throw errors, the first error that was encountered will be the result.

If you want the wait Future to complete as soon as any of the Futures throws an error, without waiting for the remaining ones to complete, you can set eagerError to true in the wait constructor. This can significantly improve efficiency when you need multiple Futures to complete before proceeding, but they do not need to wait for each other.

There may also be a scenario where you can start processing the result of a Future before the others have completed. Let's say a user selects three categories in the search menu and we need to show them search results for all categories. This can be better understood through the following code example:

lib/examples/example14.dart

```
void main() async {
  final sw = Stopwatch()..start();
  final searchResult = await search('candy');
  final mapResult = await map(searchResult);
  print(mapResult);
  final searchResult2 = await search('chocolate');
  final mapResult2 = await map(searchResult2);
  print(mapResult2);
  final searchResult3 = await search('ice cream');
  final mapResult3 = await map(searchResult3);
  print(mapResult3);
  sw.stop();
  print('Total time: ${sw.elapsedMilliseconds}ms');
}

Future<String> search(String query) async {
  // Simulate network delay
  await Future.delayed(Duration(seconds: 3));
  return 'Search results for $query';
}

Future<String> map(String query) async {
  await Future.delayed(Duration(seconds: 3));
  return 'Mapped: $query';
}
```

In this code, we perform a `search` operation for an element for 3 seconds, and then we `map` the result of that search for another 3 seconds. We repeat this process three times for each query, resulting in a total operation time of approximately 18 seconds. However, the independent search results do not need to wait for each other, even though the `map` functions depend on the `search` function's result. So, let's optimize this code:

lib/examples/example15.dart

```
void main() async {
  final sw = Stopwatch()..start();
  final searchResultFuture = search('candy');
```

```
    final searchResultFuture2 = search('chocolate');
    final searchResultFuture3 = search('ice cream');

    final mapResultFuture = map(await searchResultFuture);
    final mapResultFuture2 = map(await searchResultFuture2);
    final mapResultFuture3 = map(await searchResultFuture3);

    final results = await Future.wait([
      mapResultFuture,
      mapResultFuture2,
      mapResultFuture3,
    ]);

    print(results);
    sw.stop();
    print('Total time: ${sw.elapsedMilliseconds}ms');
}
```

Now, when we run this code, we will achieve the same result in approximately 6 seconds. This is three times faster than before! Let's review the steps we took to accomplish this:

1. We created a Future for each search result we needed. Remember, the Future is posted on the event loop at the time of creation, so it starts executing immediately, even without `await`.

2. Since we depend on the search results for our `map` functions, we need to await the results. So, we created futures for our `map` results. While we await the first `search`, the second and third are executed simultaneously, saving us time.

3. Finally, we kickstarted three of our `map` functions and now we just need to wait for them. Since the `map` functions are independent, we don't need to await each one individually. Instead, we can use `Future.wait` to kickstart all of them.

This demonstrates how understanding the inner workings of an API – in this case, the Future API – helps us structure our code efficiently and optimally. It is crucial to note here that this technique is very use case dependent – be mindful of potential race conditions if the independent methods modify shared mutable resources. While this approach works well for some cases, there are situations where a single operation may still take too much time, leading to frame drops. In Flutter, when the UI stutters, it's a sign of frame dropping, indicating that the main thread is overloaded and needs to be offloaded. This is where a new concept called **isolates** comes to the rescue.

But before we continue exploring isolates, our discussion of concurrency in Dart would be incomplete without mentioning the Streams API. In *Chapter 4*, we saw how Streams is the Dart implementation of the observable pattern, as well as learned how to work with `StreamController`. In essence, where the `Future` class returns a single value, `Stream` can emit multiple values over the duration.

You can refer to the example in *Chapter 4* in the *Emitting data via Streams* section to refresh your memory. Now, let's explore how we can achieve parallelism in Dart with the help of isolates.

Embracing parallelism with isolates

Product search is one of the most essential features of a regular e-commerce app. However, a common problem that users face when searching for something is making typos. This issue is so prevalent that several algorithms are designed to solve it by returning results that closely match the search query, even if it's not an exact match. This approach is often referred to as **fuzzy search**. Algorithms like those used for implementing fuzzy search can be greedy and spend many resources on number crunching, resulting in degraded UI performance. Let's implement the fuzzy search feature in our Candy Store app and make sure our app performance is not affected with the help of isolates.

Implementing fuzzy search in the Candy Store app

To implement this feature, we will perform the following steps:

1. Add a search bar to the main products screen.
2. Implement a `List<Product> search(String query)` function in `AppProductRepository`.
3. Incorporate the Levenshtein Distance algorithm within this function.

The **Levenshtein Distance** algorithm compares two strings and calculates the number of edits required to transform the query string into the target string. These edits can include additions, deletions, and substitutions. For example, in the case of "chocalate," the Levenshtein Distance is only 1, representing the need to swap "a" for "o" to form "chocolate." We won't delve into the algorithm's implementation details as it is irrelevant to our current topic (you can refer to the full implementation in the source code).

However, there's a problem. When we run our app and perform a search, we notice that our UI stutters. Instead of a smooth experience with a progress bar and uninterrupted typing, we observe the flickering type indicator and janky text. Why does this occur?

> **Note**
> You can see an example of how it looks at `https://github.com/PacktPublishing/Flutter-Design-Patterns-and-Best-Practices/blob/master/CH09/final/candy_store/example_frame_drop.gif` or just run the code that we will discuss in the next part.

The issue lies in the fact that both the progress bar and the text field where we enter our query are being rebuilt on the screen. The progress bar relies on animation ticks, occurring 60 or 120 times per second depending on the device's refresh rate. Meanwhile, the text field is rebuilt each time we type. The problem arises when we enter a symbol and trigger the `search` function, which takes a significant amount of time. As a result, the request to rebuild our widget is delayed, leading to dropped frames and UI stuttering.

Using a `Future` class alone is insufficient to solve this problem because the `search` event takes too long to execute on the event loop. In cases involving intense number crunching, data processing, and other computationally heavy tasks, we can utilize isolates.

Understanding the concept of isolates

Strictly speaking, Dart is a single-threaded language. However, it has a concept called **isolate** that allows for not only asynchronous operations but also actual parallel operations. The main difference between isolate and a regular thread is that isolates do not share memory. This eliminates some of the common multi-threading problems such as data races (`https://dart.dev/language/concurrency#isolates`). However, this also means that isolates cannot communicate with each other via shared state and can only exchange messages through special ports. If you're curious, you can read more about this concurrency model, called the **Actor model** (`https://en.wikipedia.org/wiki/Actor_model`), which the isolates concept is based on. While this approach has many benefits, it can make the API somewhat cumbersome.

There are two main types of use cases for isolates:

- **One-off operations**: You launch an isolate, or "spawn" it in Dart terminology, it performs its job in parallel to your running application code, returns the result, and then closes.

- **Long-living isolates**: Spawning an isolate can be resource-intensive, so for scenarios where one isolate can be used for recurring operations or when inter-isolate communication is required, instantiating and maintaining long-living isolates may be preferable. This is an advanced scenario that involves intricacies and specific error handling. If this is your use case, it is recommended that you refer to the official Flutter documentation, which provides a comprehensive tutorial on this specific scenario: *Sending Multiple Messages Between Isolates with Ports* (`https://dart.dev/language/isolates#sending-multiple-messages-between-isolates-with-ports`).

Let's refactor our fuzzy search implementation so that it uses isolates and offloads the heavy computations of the Levenshtein Distance algorithm to a parallel process. Currently, our `searchProducts` function in `AppProductRepository` looks like this:

lib/product/data/repository/app_product_repository.dart

```
@override
  Future<List<Product>> searchProducts(String query) async {
    final allProducts = fakeSearchData;
    if (query.isEmpty) {
```

```
        return allProducts;
    }
    final results = _search(query);
    return results;
}

List<Product> _search(String query) {
    // Resource-consuming implementation of the Levenshtein Distance
        algorithm here
}
```

The problem is that _search is a simple function, but it takes too much time and results in frame drops. Fortunately, the API for one-off operations with isolates is straightforward. All we need to do is wrap our costly method call in an Isolate.run function, which handles all the boilerplate for setting up isolates. Let's see it in action:

lib/product/data/repository/app_product_repository.dart

```
@override
  Future<List<Product>> searchProducts(String query) async {
    final allProducts = fakeSearchData;
    if (query.isEmpty) {
      return allProducts;
    }
    final results = await Isolate.run(() => _search(query));
    return results;
}
```

By wrapping our _search function in an isolate and launching it, we encounter an error. It should look similar to the following:

```
Unhandled Exception: Invalid argument(s): Illegal argument in
isolate message: object is unsendable - Library:'dart:async' Class:
_Future@4048458 (see restrictions listed at `SendPort.send()`
documentation for more information)
```

The reason for this exception is the _search function itself. Remember that an isolate doesn't share memory with other isolates. Your entire Flutter application runs in a single isolate, hence the "Dart is single-threaded" saying. When you spawn new isolates, they need to serialize the entire context of the function used in an isolate. In our case, the _search function belongs to the AppProductRepository class. So, when we pass it to Isolate.run, we also tell it to serialize the entire AppProductRepository class. Because the class has various dependencies, we eventually run into problems. To avoid this, passing either a static or a top-level function to the isolate is recommended. So, we can add a `static` prefix to our `_search` function like this:

lib/product/data/repository/app_product_repository.dart

```
static List<Product> _search(String query) {
  // Resource-consuming implementation of the Levenshtein Distance
    algorithm here
}
```

Now, because it's static and belongs to the class rather than an instance of it, it doesn't require any dependencies. It is serialized without problems, and we can see that our search works seamlessly, our UI doesn't stutter, and everything is smooth.

One last thing to note is that isolates are not supported on the web. So, if you want to avoid errors on the web, instead of using the `Isolate.run` function, you can use the special `compute` function. On native platforms, it will use `Isolate.run`, and on the web, it will run code synchronously. The updated code would look like this:

lib/product/data/repository/app_product_repository.dart

```
@override
  Future<List<Product>> searchProducts(String query) async {
    final allProducts = fakeSearchData;
    if (query.isEmpty) {
      return allProducts;
    }
    final results = await compute(_search, query);
    return results;
  }
}
```

Here, we pass our function as the first parameter to the `compute` function and then pass `query` as the second parameter to the `_search` function.

Summary

In this chapter, we learned about a complex yet very important concept in modern-day application programming – concurrent programming and its peculiarities in Dart. We explored the inner workings of the asynchronous API in Dart called Future, as well as learned practical tips and tricks on how to write asynchronous code efficiently. We also observed various approaches to error handling when dealing with Futures. Number crunching, background tasks, serializing big chunks of data – all of these actions can take so much time that they will end up blocking the main thread, inevitably resulting in degraded app performance. Thankfully, you now know how to solve problems when asynchronicity is not enough by using parallelism via isolates and the convenient `compute` function.

In the next chapter, we will learn best practices for another set of crucial tasks in app development with Flutter – how to communicate with the underlying native platforms efficiently.

Get this book's PDF version and more

Scan the QR code (or go to `packtpub.com/unlock`). Search for this book by name, confirm the edition, and then follow the steps on the page.

Note: Keep your invoice handy. Purchases made directly from Packt don't require an invoice.

10

A Bridge to the Native Side of Development

The number one reason to use Flutter is its cross-platform capabilities. Flutter's cross-platform utility is *so* great that it's easy to forget about its connection with the underlying native platforms. For most of your Flutter development, you probably won't even need to touch a line of native code, but eventually, you might encounter a use case that will require going down to the native layer. It could be a third-party SDK that doesn't have an official Flutter plugin or a new feature available on a new version of the native OS that you're just too eager to try out even before it gets official support.

In this chapter, we will overview the architecture of a Flutter application and how it connects to the native layer. We will learn how to access this layer with the tools available out of the box, as well as explore a better way to implement this functionality.

We will cover the following main topics:

- Understanding Flutter app architecture
- Working with platform channels
- Ensuring type-safe communication via `pigeon` code generation

Technical requirements

In order to proceed with this chapter, you will need the following:

- The code from the previous chapter, which can be found here: `https://github.com/PacktPublishing/Flutter-Design-Patterns-and-Best-Practices/tree/master/CH09/final/candy_store`
- Libraries from `pub.dev` that we will connect to our application: `build_runner` (`https://pub.dev/packages/build_runner`) and `pigeon` (`https://pub.dev/packages/pigeon`)

- You will find all of the code required for this chapter here:

  - Start of the chapter: `https://github.com/PacktPublishing/Flutter-Design-Patterns-and-Best-Practices/tree/master/CH10/initial/candy_store`

  - End of the chapter: `https://github.com/PacktPublishing/Flutter-Design-Patterns-and-Best-Practices/tree/master/CH10/final/candy_store` (you can review the step-by-step refactoring in the commit history of this branch)

Understanding Flutter app architecture

The Flutter framework is composed of multiple layers, each with its own responsibility for a specific part of the application. It is important to remember that cross-platform applications are still built on top of the native layer. While Flutter and its package ecosystem do a great job of abstracting this native layer, there are situations where we still need to interact with it. Before exploring these use cases, let's take a general overview of how Flutter connects with the native layer.

Diving into the Flutter framework layers

You may have seen a diagram similar to this before; for example, in the official documentation:

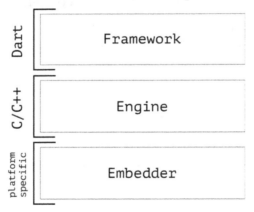

Figure 10.1 – The Flutter SDK layers

While we don't need to understand all the intricacies of each layer, it is helpful to know their responsibilities and how developers interact with them.

The layer that you, as a Flutter developer, use every day when developing your apps is the **Framework layer**. It contains all the code required to write a Flutter app, including **Material** and **Cupertino** styling, the widget system, gesture handling, and everything discussed in previous chapters, such as `ChangeNotifier` classes. The framework is written in the Dart language, so we also use Dart in the Flutter portion of our application.

The next layer is the **Engine layer**. As a developer, you don't interact with its APIs in raw format, but it provides foundational support for the Flutter framework. It is platform-agnostic and implements low-level Flutter core APIs.

The last layer in this diagram is the **Embedder** layer. This acts as the glue layer between the Flutter framework and the underlying native platform. It handles the setup required for actual rendering, event loops, and more. It exposes the `Embedder` API that the embedding platform uses to coordinate with the Flutter framework and the app.

This layered architecture ensures consistency in the APIs available to developers and the behavior on end devices. It also allows for the flexibility to accommodate the platform-specific needs of the embedder, enabling different embedders and target platforms for Flutter developers. So far, we have covered the framework part, but there are a couple more layers to consider – our actual application. Let's examine the extended diagram:

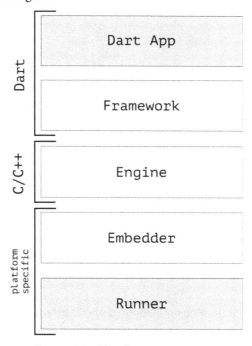

Figure 10.2 – The Flutter app anatomy

Here, we can see that there are two more layers involved: the topmost layer, the Dart application, is the part that most Flutter developers work with and create. This is the part that we have been working on in this book so far. However, there is also a foundation part: the **Runner**, or the native part of the application. As mentioned earlier, Flutter is an abstraction over the native platform, and now we can visually see this concept. Under the hood, it still compiles to a native application and often uses native APIs, especially for accessing various native features such as hardware and OS-specific functionality.

Occasionally, as developers, we may need to work with this native layer too. For example, if there is no existing package that supports the native functionality we require, such as audio, video, camera, or background work, we may need to delve into this layer. Or, we may need to perform highly specific tasks on the native side of things, using an SDK that doesn't have a Flutter API. Although such use cases may be rare, it's good to know how to handle them. Let's explore how we, as Flutter developers, can communicate with the native layer using platform channels.

How does Flutter communicate with the native layer?

A Flutter application consists of two parts: the Flutter part, referred to as the *client*, and the native part, known as the *host*. The communication between these parts is facilitated by an API called the **platform channel**. There are various types of platform channels, which we will review shortly. Schematically, it looks like this:

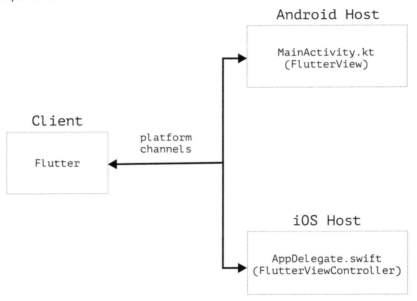

Figure 10.3 – A diagram showing the platform channels between the client and host

But a logical question arises: the client part is written in Dart, yet the native part can be written in many languages, such as Kotlin and Swift (when it comes to mobile platforms). How is interoperability ensured between these languages?

Encoding data with MessageCodec

To transform a Dart `int` type into a Kotlin `Integer` type, it must first be encoded into binary format on the Dart side and then decoded from binary format on the Kotlin side (or Swift, or C++, and so on) and vice versa. This process is facilitated by a concept called `MessageCodec` in Flutter platform channels, with one of the default implementations being `StandardMessageCodec`. The `StandardMessageCodec` implementation implements two methods of `MessageCodec`: `encodeMessage` and `decodeMessage`, as well as some additional methods for supporting basic types. It supports a limited set of types, which are listed in the following table (`https://docs.flutter.dev/platform-integration/platform-channels?tab=type-mappings-kotlin`):

Dart	Kotlin	Swift
Null	null	nil
Bool	Boolean	NSNumber(value: Bool)
Int	Int	NSNumber(value: Int32)
int (if 32 bits is not enough)	Long	NSNumber(value: Int)
double	Double	NSNumber(value: Double)
String	String	String
Uint8List	ByteArray	FlutterStandardTypedData(bytes: Data)
Int32List	IntArray	FlutterStandardTypedData(int32: Data)
Int64List	LongArray	FlutterStandardTypedData(int64: Data)
Float32List	FloatArray	FlutterStandardTypedData(float32: Data)
Float64List	DoubleArray	FlutterStandardTypedData(float64: Data)
List	List	Array
Map	HashMap	Dictionary

Table 10.1 – List of supported types by StandardMessageCodec

Basically, you can pass around numbers, strings, Booleans, and various collections. Additionally, you have the option to use a different codec, such as `JSONMessageCodec`, or even create your own encoding and decoding extensions if you need to handle complex custom objects.

To communicate between the Flutter (client) and native (host) parts of the app or vice versa, you use an API called platform channels. These channels enable the exchange of different data types, supported by various codecs, such as `StandardMessageCodec`.

To summarize, the following diagram illustrates this process:

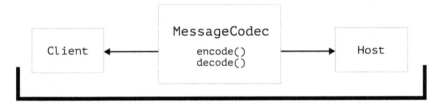

Figure 10.4 – Message encoding between platform channels

Now that we understand the theoretical aspects of this communication, let's learn how to put it into practice.

Working with platform channels

We will be implementing a new feature in our `candy_store` app. The feature is a **Favorites** page that will display a list of items that we have favorited within the app. This list will be saved and persist between app restarts. There are various ways to store different types of data on a device, but the method depends on the operating system being used. For example, in native iOS, we can use `UserDefaults` to store basic key-value data, while in native Android, we have `SharedPreferences` for that purpose. Fortunately, Flutter provides an official plugin called `shared_preferences` (https://pub.dev/packages/shared_preferences) that abstracts these APIs and offers a convenient Dart interface. This plugin is typically the recommended choice for handling basic data storage. However, for the purpose of demonstrating how to work with platform channels, we will create our own implementation that is tailored to our application. Let's begin by selecting the appropriate channel type for this task.

Selecting the platform channel type

So far, we have discussed the concept of platform channels, which is an umbrella term for several specific types. The most common types are `BasicMessageChannel`, `MethodChannel`, and `EventChannel`. Each type of channel is used for different purposes and has its own characteristics. Let's take a closer look at them:

- The `BasicMessageChannel` type is the low-level API platform channel. It allows sending simple "messages," which are encoded using the aforementioned `StandardMessageCodec` implementation.

- The `MethodChannel` type is a more popular type of channel. It provides a convenient API for calling specific methods by supplying a method name and an optional list of arguments. By default, it uses `StandardMethodCodec`, which supports the same types as `StandardMessageCodec` but also includes some method-specific code.

- The `EventChannel` type is a specialized channel used when you want to continuously stream data from the native side to the Flutter side. It is commonly used for scenarios such as receiving data from sensors. It allows opening a connection between the client and the host and, on the Dart side, it is exposed as a Dart Stream with various useful APIs.

For our use case, `BasicMessageChannel` is a bit too low-level, and we don't need to stream data from the native side; therefore, `EventChannel` is not suitable either. This leaves us with `MethodChannel`, which offers exactly what we need. We can use it to call methods that add an item to **Favorites**, remove an item from **Favorites**, or retrieve a list of already favorite items. So, let's use it!

Implementing the UI of the Favorites feature

We will start with the code available here (`https://github.com/PacktPublishing/Flutter-Design-Patterns-and-Best-Practices/tree/master/CH10/initial/candy_store`) and step by step will arrive at the final version available here (`https://github.com/PacktPublishing/Flutter-Design-Patterns-and-Best-Practices/tree/master/CH10/final/candy_store`). You can review the commit history to see all of the steps. To implement this feature in its initial status, we have applied everything we have learned so far. First, let's take a look at how it works:

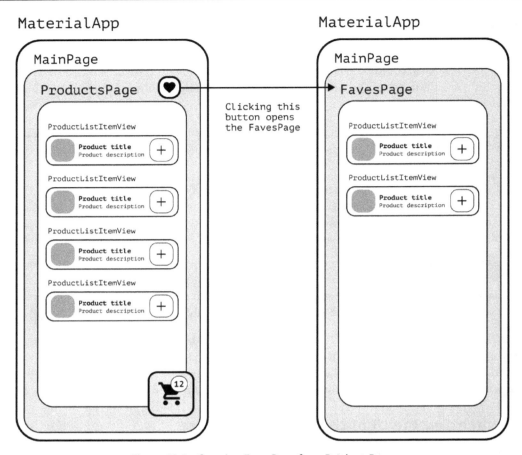

Figure 10.5 – Opening FavesPage from ProductsPage

From the initial `ProductsPage` page, we can open the `FavesPage` page by clicking on the heart icon in the `AppBar` component. On the `FavesPage` page, we see a list of items that we have added to **Favorites**. We fetch those items from an abstract class called `FavesRepository` that has methods for addition, removal, fetching, and checking if an item is a favorite. Its interface looks like this:

lib/faves/domain/repository/faves_repository.dart

```
abstract class FavesRepository {
  Future<List<ProductListItem>> getFaves();
  Future<void> addFave(ProductListItem item);
  Future<void> removeFave(String id);
  Future<bool> isFave(String id);
}
```

Before implementing persistent storage, we will temporarily store all the information in memory. However, this data will be lost once the user closes our application. This implementation is straightforward, as we just maintain a local list called _faves:

lib/faves/data/repository/in_memory_faves_repository.dart

```dart
class InMemoryFavesRepository implements FavesRepository {
  final List<ProductListItem> _faves = [];

  @override
  Future<void> addFave(ProductListItem item) async {
    _faves.add(item);
  }

  @override
  Future<List<ProductListItem>> getFaves() async {
    return _faves;
  }

  @override
  Future<void> removeFave(String id) async {
    _faves.removeWhere((element) => element.id == id);
  }

  @override
  Future<bool> isFave(String id) async {
    return _faves.any((element) => element.id == id);
  }
}
```

In addition to viewing favorite items, users can also add and remove them. This functionality is available from the app bar of ProductDetailsPage. The business logic is implemented using Bloc, in the same way we did in *Chapter 4*.

However, having favorites that are cleared when the user closes the application is kind of useless. We want users to be able to access their favorite items at any time. To solve this issue, we will use the native functionality of SharedPreferences on the Android side and UserDefaults on the iOS side, as discussed previously. Let's implement it!

Using MethodChannel for favoriting items

Let's begin by creating a new repository. Later on, we can replace it via **dependency injection (DI)** in a single place, and everything will work out of the box. This again demonstrates the beauty of the repository and DI patterns discussed in *Chapters 6* and *7*.

We will name it `LocalFavesRepository` since it will store favorites locally on the device rather than remotely on a backend:

lib/faves/data/repository/local_faves_repository.dart

```
class LocalFavesRepository implements FavesRepository {
  static const _platform =
  MethodChannel('com.example.candy_store/faves');
}
```

Besides creating the new repository, we have also added a field called `_platform`. This field is used to create a `MethodChannel` channel on the Flutter side. When creating a `MethodChannel` channel, it is important to provide a unique name for the channel in the constructor. This uniqueness applies not only to the channels you create personally but also to those created by libraries and other components. Using a non-unique channel name can result in inconsistent and buggy behavior. To ensure uniqueness, a good practice is to prefix the channel name with your package name. In our case, the package name is `com.example.candy_store`, and we add `/faves` to it to form the channel name.

Next, we need to establish a connection to this channel on the native side. To do this, in the `MainActivity.kt` file from the Android part of the app, we add the following code:

android/app/src/main/kotlin/com/example/candy_store/MainActivity.kt

```
class MainActivity : FlutterActivity() {

    override fun configureFlutterEngine(flutterEngine: FlutterEngine){
        super.configureFlutterEngine(flutterEngine)
        val favesMethodChannel = MethodChannel(
            flutterEngine.dartExecutor.binaryMessenger,
            "com.example.candy_store/faves"
        )
    }
}
```

Before we dive into an explanation of `MethodChannel` channels on the Android side, let's make sure that there is no confusion related to Kotlin.

Understanding Kotlin

Dart's syntax is similar to many languages, especially Kotlin and Swift, so you might intuitively understand the code in the `MainActivity.kt` file as it is. But to make it clearer, let's gloss over some Kotlin syntax from this code snippet:

- In order to override a method in Dart, we would use an `@override` annotation, while in Kotlin, it's an `override` function modifier.

- In Dart, we don't need to use any special keywords that declare a function, but in Kotlin, we need to use the `fun` modifier.

- In Dart, in the function argument list, we first specify the type and then the name – for example, `configureFlutterEngine(FlutterEngine flutterEngine)` – but in Kotlin, it's the other way around; first the name, followed by : and the type, like this: `configureFlutterEngine(flutterEngine: FlutterEngine)`.

- In Dart, we use the `final` keyword to initialize a variable that cannot change its value after initialization. In Kotlin, we use the `val` keyword for that.

Now, let's dive into what's going on in the `MainActivity.kt` file. There are a couple of things to unwrap here:

1. To connect a `MethodChannel` channel on the Android side, we need to override the `configureFlutterEngine` method.

2. This is necessary so that we can access `binaryMessenger` from the Flutter engine. This messenger is responsible for facilitating communication across platform channels. In most cases, you will be using the default messenger. If you ever need a custom one, it is beyond the scope of this book.

3. We also need to pass the channel name in the constructor. It is crucial for this channel name to exactly match the one on the Flutter side. Otherwise, you will encounter errors, which we will discuss soon.

Great – we have set up our platform channel! Now, let's actually call some methods. On the Flutter side, in order to add an item to **Favorites** by passing the ID of the item in our `LocalFavesRepository` repository, we would write this code:

lib/faves/data/repository/local_faves_repository.dart

```
@override
  Future<void> addFave(ProductListItem item) async {
    _platform.invokeMethod(
      'addFave',
      {'id': item.id},
```

```
    );
  }
```

Let's review what we have just done:

1. As the first parameter to the `invokeMethod` method of `MethodChannel`, we pass `addFave`, a string that represents the name of the method to be called on the native side.

2. To pass parameters, we create a map and add an `item.id` single value for the `'id'` key.

Now, on the native Android side, back in our `configureFlutterEngine` method, we will write this code:

android/app/src/main/kotlin/com/example/candy_store/MainActivity.kt

```kotlin
override fun configureFlutterEngine(flutterEngine: FlutterEngine) {
    super.configureFlutterEngine(flutterEngine)
    val favesMethodChannel = MethodChannel(
        flutterEngine.dartExecutor.binaryMessenger,
        "com.example.candy_store/faves"
    ) // #1
    favesMethodChannel.setMethodCallHandler {call, result ->//#2
        when (call.method) { //#3
            "addFave" -> { // #4
                val id = call.argument<String>("id") // #5
                if (id == null) {
                    result.error(
                    "INVALID_VALUE",
                    "id is null",
                    null) // #6
                } else {
                    toggleFavorite(id, true)
                    result.success(null) // #7
                }
            }

            else -> {
                result.notImplemented() // #8
            }
        }
    }
}
```

Now, we have even more things to unwrap here! Here's what we've done:

1. We have added a method handler to the `favesMethodChannel` method channel using the `setMethodCallHandler` method.

2. When we override this method, we receive two parameters – `call` and `result`. The `call` parameter contains information about the method called from the Flutter side, including the method name and arguments. The `result` parameter is used to pass information back to the Flutter side after we have finished processing the method.

3. Since we can have multiple methods passed through a single `MethodChannel` channel, we need a way to identify which specific method was called. To achieve this, we check the `call.method` parameter.

4. In this case, we are implementing the `addFave` method and handling it here. Similarly, we would add handlers for other method names. It is crucial that the method names match exactly, as we are performing string comparisons.

5. Now, we need to retrieve the parameters from the `call` parameter. It's important to note that these parameters can be nullable. Since we can't ensure strong typing and guarantee that the caller will pass a parameter named `id` of type `String`, we need to perform null checking and type casting. This process can be inconvenient and prone to errors. We will explore ways to improve it soon. For now, this is the default approach.

6. If the parameter is `null`, which is an invalid case for us, we should return an error to the caller. We can do this by calling `result.error` and passing the error parameters. This notifies the caller that the method execution has completed with an error.

7. On the other hand, if everything goes well, we should call `result.success`. Depending on the method, we may want to return a result. For example, in the case of `getFaves`, it could be a `List<String>` type containing `id` parameters. In the case of `addFave`, we don't have anything to return, so we simply pass `null`, which basically means void. It's important to return a result in some form because otherwise, the asynchronous method on the Flutter side will never complete.

8. This brings us to the last step. If we receive a method name that we don't know how to handle, we still need to return a result. In this case, we call `result.notImplemented`, which completes with a specific error on the Flutter side.

Inter-platform communication is asynchronous in nature. In Dart, this means using `Future` objects when awaiting a method call from the native side. To ensure that the `Future` object completes, we use the `Result` class with its methods: `success`, `error`, or `notImplemented`. It is crucial to remember to return a result, since failing to do so will cause the `Future` object to hang indefinitely, which is not something we want. Additionally, it is important to only return a single result. Calling multiple methods on the same result will throw an `IllegalStateException` exception. This mistake is easy to make if your code contains numerous conditionals and logic flows.

Arguably, this isn't the most convenient API as it enables developers to create bugs quite easily. Another important consideration is that invoking channel methods should be done on the platform's main thread. Otherwise, it will lead to errors and unexpected behavior.

The iOS side conceptually looks very similar. Let's review the details:

ios/Runner/AppDelegate.swift

```swift
@objc class AppDelegate: FlutterAppDelegate {

    override func application(
        _ application: UIApplication,
        didFinishLaunchingWithOptions launchOptions
            : [UIApplication.LaunchOptionsKey: Any]?
            ) -> Bool {
        // #1
        let controller: FlutterViewController
            = window?.rootViewController as! FlutterViewController
        // #2
        let favesChannel = FlutterMethodChannel(
            name: "com.example.candy_store/faves",
            binaryMessenger: controller.binaryMessenger
        )
        // #3
        favesChannel.setMethodCallHandler { [weak self] (
            call: FlutterMethodCall,
            result: @escaping FlutterResult) in guard let self =
            self else { return }
            switch call.method {
            case "getFaves":
                result(self.getFaves())
            case "addFave":
                // #4
                if let args = call.arguments as? [String: Any],
                    let id = args["id"] as? String {
                    self.toggleFavorite(id, isFavorite: true)
                    result(nil)
                } else {
                    result(
                        FlutterError(
                            code: "INVALID_VALUE",
                            message: "id is null",
                            details: nil)
                    )
```

```
        }
        // ... other methods
    default:
        result(FlutterMethodNotImplemented) // #5
    }
  }
  }
}
```

Let's look at what we've just done:

1. First, we have obtained the `FlutterViewController` controller in order to access `binaryMessenger`.

2. Next, we have created an instance of `MethodChannel`. Once again, the name should completely match the one defined on the Flutter side.

3. After that, we set the method handler on the channel, where we match strings as method names, as before.

4. Similar to Kotlin, we have unwrapped potentially nullable arguments and implemented error handling for invalid argument cases.

5. Finally, we have returned a result, which can be `nil` or values (such as in `getFaves`) indicating success, `FlutterError` in case of an error, or `FlutterMethodNotImplemented` if we don't know how to handle the method.

Some of this code may be new to you, so let's look at the Swift code in more detail.

Understanding Swift

You might already intuitively understand the code in the `AppDelegate.swift` file as it is, but to make it clearer, let's gloss over some Swift syntax from this code snippet:

- In order to override a method in Dart, we would use an annotation `@override` annotation, while in Swift, it's a function modifier called `override`.

- In Dart, we don't need to use any special keywords that declare a function, but in Swift, we need to use the `func` modifier.

- In Dart, in the variable initialization, we first specify the type, and then the name, (for example, `FlutterViewController controller`), but in Swift, it's the other way around – first the name, followed by : and the type, like this `controller: FlutterViewController`.

- In Dart, we use the `final` keyword to initialize a variable that cannot change its value after initialization. In Swift, we use the `let` keyword for that.

- In Swift, `nil` means the same as `null` in Dart.

- In Swift, we can use Any for a variable that could be of any type, similar to the Dart dynamic type. This piece of code utilizes Swift functionality called optional binding:

```
if let args = call.arguments as? [String: Any],
    let id = args["id"] as? String {
    self.toggleFavorite(id, isFavorite: true)
    result(nil)
} else {
    result(
        FlutterError(
            code: "INVALID_VALUE",
            message: "id is null",
            details: nil)
    )
}
```

What this means is that if we can safely cast call.arguments to be a map of type [String: Any], and this map contains a String value with the "id" key, then execute the following block of code. Otherwise, execute the else code block.

What we have seen so far is a quite simple example, yet even at this point, many potential problems arise. Let's review what they are.

Overviewing the problems introduced by vanilla MethodChannel channels

There are many potential problems and inconveniences when working with MethodChannel channels:

- Channel names must match across all supported platforms. Making a typo or forgetting to rename a channel will result in a MissingPluginException exception being thrown on the Flutter side. The same exception will be thrown if a method is not handled, such as not returning result.notImplemented for an unexpected method name.

- Method parameters are simple Map types with dynamic parameters. This requires a lot of type casting and error checking. Additionally, if there are any typos in the parameter names, a PlatformException exception will be thrown.

- Working with the Result API requires caution. If we don't return a reply using one of the result methods, the Future object on the Flutter side will not complete. Furthermore, if we call multiple methods on the same result, an exception will be thrown – IllegalStateException("Reply already submitted").

So, to sum up our experience, we have encountered the possibility of exceptions being thrown at every step, along with a lot of string matching, type casting, and error handling. As the features we work on become more complex and require a lot of back-and-forth communication, this process becomes even more tedious and error-prone. Fortunately, there is a tool that can help us address these problems and improve the experience of working with the native layer. That tool is `pigeon`, a code generation package. Let's explore how `pigeon` works and how it resolves these issues.

Ensuring type-safe communication via pigeon code generation

Working with Flutter means we're working with Dart on the Flutter side and Kotlin and Swift on the native mobile side. One common feature of these programming languages is that they are statically type-safe. In simple terms, this means that if we declare a variable `a` as type `int`, it will remain an `int` type throughout its lifetime. This allows us, as programmers, to always be certain of the variable's type and safely perform type-specific operations on it.

Moreover, this type safety is ensured during compile time, meaning that any bugs related to type mismatches are caught before running the app. By making types dynamic, we defer type checking to runtime, which leads us to all of the issues we have just discussed. The `pigeon` package solves these problems with code generation; it generates type-safe bindings for all of the required platforms based on code annotations and a specified interface. Let's see how it works in practice.

Configuring a pigeon interface

First of all, as noted in the *Technical requirements* section, we need to add `pigeon` (`https://pub.dev/packages/pigeon`) and `build_runner` (`https://pub.dev/packages/build_runner`) as dev dependencies. Next, we will create an input interface for the pigeon code generator. To do that, we need to do the following:

- Create a folder *outside of the lib folder* that will contain our pigeon interface. For example, we will call the folder `pigeon` and it will be on the same level as `lib`. Our app structure will look like this:

  ```
  candy_store
  ├── android
  ├── ios
  ├── lib
  └── pigeon
  ```

- Inside of this folder, we create a file that will be our interface. Since the feature we're implementing is working with local storage, let's call this interface `LocalStorageApi` and the file `local_storage_pigeon_api.dart`.

Now, let's move on to the coding part. In this file, we describe the interface of what we want our native communication to look like. This is simple Dart code:

pigeon/local_storage_pigeon_api.dart

```dart
import 'package:pigeon/pigeon.dart'; // #5

@HostApi() // #1
abstract class LocalStorageApi { // #2
  void addFave(String id);

  void removeFave(String id);

  bool isFave(String id);

  List<FaveProduct> getFaves(); // #3
}

class FaveProduct { // #4
  final String id;

  FaveProduct(this.id);
}
```

There are several interesting things to note here:

1. As mentioned before, `pigeon` operates based on code annotations. This means that as developers, we annotate the code that we want to generate and then use the code generation tool to generate code based on these annotations. In our case, we are creating an interface for calling code from the client (Flutter) to the host (native), which is why we annotate our interface with the `@HostApi()` annotation.

2. After that, we create a regular Dart interface, just like we would in a typical Flutter app.

3. This is where the peculiarities of `pigeon` come into play. The current limitation of `pigeon` (as of version 12.0) is that we cannot import any files or libraries into this file; otherwise, the generator will throw an error. So, if we want to use any custom classes that are not already present in Dart, we need to define them within this exact file.

4. Let's consider our `FaveProduct` class as an example of a complex class. Since it is a custom class, we define it in the `locale_storage_pigeon_api.dart` file. This may not be convenient, especially if we have many custom classes, but it is a trade-off for the sake of type safety. In my opinion, the benefits of type safety outweigh this inconvenience by a large margin.

5. Additionally, it is worth mentioning that the only allowed import in this file is `import 'package:pigeon/pigeon.dart';`, which allows us to use the annotations. If any other `import` statements are added, the generator won't work.

So far, we have described the desired structure of our API using regular Dart code. Before generating the code, we need to take one more step, which is to specify the names of the output files that we want to generate. Here is how we can do this:

pigeon/local_storage_pigeon_api.dart

```
@ConfigurePigeon(
  PigeonOptions(
    dartOut: 'lib/faves/data/api/local_storage_api.g.dart',
    kotlinOut:
        'android/app/src/main/kotlin/
        'com/example/candy_store/LocalStorageApi.g.kt',
    kotlinOptions: KotlinOptions(package: 'com.example.candy_store'),
    swiftOut: 'ios/Runner/LocalStorageApi.g.swift',
    dartPackageName: 'candy_store',
  ),
)
@HostApi()
abstract class LocalStorageApi { // same code as before }
```

We add another annotation, `@ConfigurePigeon`, and pass an object of type `PigeonOptions` as a parameter. This object contains straightforward parameters: the file paths for Dart, Kotlin, and Swift files where the tool should generate code. A helpful tip is to suffix the filenames with g, making it easy to identify that the code was generated. For example, in `swiftOut`, the filename is `LocalStorageApi.g.swift`.

Now that our configuration is in place, we can generate code based on it. To do this, we run the following command:

```
flutter pub run pigeon --input pigeon/local_storage_pigeon_api.dart
```

We run this command from the root of our project, and as an `--input` parameter, we pass the path to where the interface we described is located. Once this command is executed, three new files are generated (based on the `dartOut`, `kotlinOut`, and `swiftOut` paths). The simplified version of our app structure now looks like this:

```
candy_store
├── android
│   ├── MainActivity.kt
│   └── LocalStorageApi.g.kt #2
├── ios
│   ├── AppDelegate.swift
│   └── LocalStorageApi.g.swift #2
├── lib
```

```
|      └── local_storage_api.g.dart #2
└── pigeon
       └── local_storage_pigeon_api.dart #1
```

So, to recap, in #1, we have defined an interface for pigeon. Next, we run a code generation command that generates three files based on the interface described in #1. These files are referenced as #2 in the file structure. But what are these files, and how do they replace the MethodChannel channels? Let's start with exploring the Dart side of things to gain a better understanding.

Connecting client and host via pigeon-generated code

Based on our interface, a new class named LocalStorageApi was generated in local_storage_api.g.dart. We can now use this class in our code. It consists of generated boilerplate code that sets up the platform channel communication for us, eliminating the need for us to do it ourselves. Before we dive into the details, let's see how we can use it. Previously, in our LocalFavesRepository repository, we had code like this:

lib/faves/data/repository/local_faves_repository.dart

```
class LocalFavesRepository implements FavesRepository {
  static const _platform =
  MethodChannel('com.example.candy_store/faves');

  @override
  Future<void> addFave(ProductListItem item) async {
    _platform.invokeMethod(
      'addFave',
      {'id': item.id},
    );
  }

  // ... the rest of the methods
}
```

We have used an instance of MethodChannel to call methods on the native side. But now, we can get rid of it completely and refactor our code to look like this:

lib/faves/data/repository/local_faves_repository.dart

```
import 'package:candy_store/faves/data/api/local_storage_api.g.dart';

class LocalFavesRepository implements FavesRepository {
  final LocalStorageApi _api;
```

```
  const LocalFavesRepository({
    required LocalStorageApi api,
  }) : _api = api;

  @override
  Future<void> addFave(ProductListItem item) => _api.addFave(item.id);

  // ... the rest of the methods
}
```

What we have done is replaced the `MethodChannel` channel with an object of type `LocalStorageApi` from `package:candy_store/faves/data/api/local_storage_api.g.dart`. We inject it in the constructor, similar to how we did DI in *Chapter 7*. The important thing to note is that this class was generated by `pigeon` and conforms to the interface we defined earlier. As a result, we have access to all the methods related to favorites manipulation in pure Dart, without having to deal with platform-channel logic or manipulate strings and types on the Dart side. This abstraction ensures type safety. Let's take a look at a diagram to understand the moving parts better:

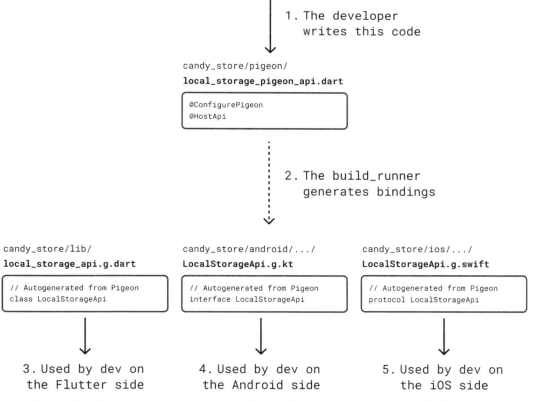

Figure 10.6 – Relationship between code written by developer and code generated by pigeon

So, this is what happens:

1. We, as developers, create a file outside of the `lib` directory (in this case, in `pigeon`), and in that file describe an interface that we would want to use. We annotate this interface with `@HostApi` annotation and specify configuration parameters via `@ConfigurePigeon`.

2. We then run the `pigeon` tool via the `flutter pub run pigeon` command, which generates three files for us (which we specified in `@ConfigurePigeon`).

3. On the Flutter side of our app, we now use the `LocalStorageApi`-generated class instead of using method channels.

4. On the Android side of our app, we now implement the `LocalStorageApi`-generated interface instead of using method channels.

5. On the iOS side of our app, we now implement the `LocalStorageApi`-generated protocol instead of using method channels.

Now, let's take a look at the generated `LocalStorageApi` class to understand how this type safety is achieved. Keep in mind: this code is autogenerated by `pigeon`, not written by hand. You will not need to write it, but it helps to understand what it does under the hood. Here is a truncated version of the code generated in `local_storage_pigeon_api.g.dart`:

lib/faves/data/api/local_storage_pigeon_api.g.dart

```
// Autogenerated from Pigeon (v18.0.0), do not edit directly.
// See also: https://pub.dev/packages/pigeon
class LocalStorageApi {
  LocalStorageApi({BinaryMessenger? binaryMessenger}) // #1
      : _binaryMessenger = binaryMessenger;

  final BinaryMessenger? _binaryMessenger;
  static const MessageCodec<Object?> codec = _LocalStorageApiCodec();
// #4

  // #2
  Future<void> addFave(String arg_id) async {
    // #3
    final BasicMessageChannel<Object?> channel =
      BasicMessageChannel<Object?>(
        'dev.flutter.pigeon.candy_store.LocalStorageApi.addFave',
        codec,
        binaryMessenger: _binaryMessenger);
    final List<Object?>? replyList =
      await channel.send(<Object?>[arg_id]) as List<Object?>?;
    // ... error handling code omitted for demo purposes
```

```
    }
}
```

At first, this code may seem intimidating, but once we understand what's happening, we will see that there is no magic involved:

1. We are already familiar with the concept of `binaryMessenger`, so if we ever want to use a custom one, we can. However, in our scenario, we will delegate it to the default one.

2. For each method described in our interface, an implementation is generated with the same signature. From the outside, it appears as regular Dart code.

3. For simplicity, `pigeon` uses `BasicMessageChannel` instead of `MethodChannel`. This doesn't make any difference to us, but it's good to know. `pigeon` generates a new dedicated channel for each described method, assigns it a name, and handles all the argument, result, and error mapping. This means that when we use the `LocalStorageApi` interface, we don't necessarily need to know that it uses native communication under the hood.

4. Finally, to support custom types such as `FaveProduct`, `pigeon` generates a custom codec that abstracts away all the encoding and decoding logic. This is better than using a JSON-based encoder because, with JSON, the object needs to be converted into a string first and then into binary. The custom codec generated by `pigeon` directly converts everything to binary, making it faster and eliminating the need for us to know anything about encoding at all.

OK – so we got rid of `MethodChannel` managing on the Dart side. Now let's get rid of it on the native side. We will start with Android in `MainActivity.kt`. Let's recall how our code used to look with `MethodChannel` channels:

android/app/src/main/kotlin/com/example/candy_store/MainActivity.kt

```kotlin
class MainActivity : FlutterActivity() {

    override fun configureFlutterEngine(flutterEngine: FlutterEngine) {
        super.configureFlutterEngine(flutterEngine)
        val favesMethodChannel = MethodChannel(
            flutterEngine.dartExecutor.binaryMessenger,
            "com.example.candy_store/faves"
        )
        favesMethodChannel.setMethodCallHandler { call, result ->
            when (call.method) {
                "addFave" -> {
                    val id = call.argument<String>("id")
                    if (id == null) {
                        result.error(
                        "INVALID_VALUE",
```

```
                                "id is null",
                                null)
                    } else {
                        toggleFavorite(id, true)
                        result.success(null)
                    }
                }
                else -> {
                    result.notImplemented()
                }
            }
        }

        private fun toggleFavorite(id: String, isFavorite: Boolean) {
            // implementation
        }
    }
}
```

We had to do everything manually. Now, `pigeon` has generated a new class for us, `LocalStorageApi`, but in Kotlin. First, let's take a look at how we can use it:

android/app/src/main/kotlin/com/example/candy_store/MainActivity.kt

```
class MainActivity:FlutterActivity(), LocalStorageApi { //#1

    override fun configureFlutterEngine(flutterEngine: FlutterEngine) {
        super.configureFlutterEngine(flutterEngine)
        // #3
        LocalStorageApi.setUp(
            flutterEngine.dartExecutor.binaryMessenger,
            this
        )
    }

    override fun addFave(id: String) { // #2
        toggleFavorite(id, true)
    }

    private fun toggleFavorite(id: String, isFavorite: Boolean) {
        // implementation
    }
}
```

Note: In Kotlin and Swift, we don't use keywords such as `extends` and `implements`. The first class specified after : is the one we extend from, and everything after , are the interfaces we implement. Hence, the `MainActivity:FlutterActivity()`, `LocalStorageApi` class can be read like `class MainActivity extends FlutterActivity implements LocalStorageApi`.

Wow! We got rid of almost all of the code. But where did it go? Here's what happened:

1. Our `MainActivity` class now implements `LocalStorageApi`, which was generated by `pigeon`. We will see the details in just a bit.

2. Instead of calling local methods and returning the result via the `Result` class, we simply implement the logic of the method in pure Kotlin, without any additional APIs, as if we were writing native code.

3. To bring everything together, we need to call the `setUp` method of `LocalStorageApi`, passing two parameters. The first parameter is `binaryMessenger`, as before, and the second parameter is `this`, referring to the `MainActivity` class that implements `LocalStorageApi`. In real-world scenarios, you would likely implement this interface in a separate class and inject it into `MainActivity`, passing the reference here. However, for simplicity's sake, we will keep it as is.

And that's it! We can stop here, and everything will work. No more `MethodChannel` channels for us to write; no string parsing for arguments and results. We have type-safe Dart and type-safe Kotlin code. However, let's take a look at what was generated on the Kotlin side by `pigeon` just to have a complete picture. Code generation tools greatly aid developers, saving time and reducing complexity, yet it's still important to understand what is happening under the hood. `pigeon` has generated a file called `LocalStorageApi.g.kt`, as we described in the `ConfigurePigeon` annotation. This file contains a Kotlin interface named `LocalStorageApi`. Here is a truncated version of it:

android/app/src/main/kotlin/com/example/candy_store/LocalStorageApi.g.kt

```kotlin
// Autogenerated from Pigeon (v18.0.0), do not edit directly.
// See also: https://pub.dev/packages/pigeon
// #1
interface LocalStorageApi {
    fun addFave(id: String)
    fun removeFave(id: String)
    fun isFave(id: String): Boolean
    fun getFaves(): List<FaveProduct>

    companion object {
        /** The codec used by LocalStorageApi. */
        val codec: MessageCodec<Any?> by lazy {
```

```kotlin
            LocalStorageApiCodec // #2
    }

    /** Sets up an instance of `LocalStorageApi`
     * to handle messages through the `binaryMessenger`. */
    @Suppress("UNCHECKED_CAST")
    // #3
    fun setUp(binaryMessenger: BinaryMessenger,
            api: LocalStorageApi?) {
        run {
            // #4
            val channel = BasicMessageChannel<Any?>(
                binaryMessenger,
                "dev.flutter.pigeon.candy_store.LocalStorageApi.
                  addFave",
                codec
            )
            if (api != null) {
                channel.setMessageHandler { message, reply ->
                    val args = message as List<Any?>
                    val idArg = args[0] as String // #5
                    var wrapped: List<Any?>
                    try { // #6
                        api.addFave(idArg)
                        wrapped = listOf<Any?>(null)
                    } catch (exception: Throwable) {
                        wrapped = wrapError(exception)
                    }
                    reply.reply(wrapped) // #7
                }
            } else {
                channel.setMessageHandler(null)
            }
        }
        // the rest of the methods
    }
  }
}
```

Once again, it's important to remember that this code is generated and not something we need to write ourselves. It is automatically generated to eliminate the need to write boilerplate code every time. Now, let's break down what's happening here:

1. We previously described and annotated a Dart interface with @HostApi. Using this interface, pigeon generates not only Dart code but also Kotlin code to ensure that we have matching interfaces on both the Dart and Kotlin sides.

2. To support binary encoding of custom types, a custom codec is also generated. This codec abstracts and maintains consistent encoding and decoding logic on both the Dart and Kotlin sides.

3. The setUp method, which we just called in our MainActivity.kt file, is generated. This method contains all code related to setting up the platform channel.

4. A matching channel is created on the Kotlin side to correspond with the one on the Dart side. The advantage is that we don't need to remember its name, as it is generated for us.

5. Type casting is handled automatically, ensuring that types match on both sides. This simplifies the process and reduces the chance of errors.

6. Compared to vanilla method channels, an important improvement is that all platform code is wrapped in a try/catch block and then wrapped again in a PlatformException exception in the catch block. This makes error handling on the Dart side much easier, as we can rely on the error always being of type PlatformException.

7. Finally, do you remember how we had to use result to finish a method before? Now, with BasicMessageChannel, we simply use reply to accomplish the same thing. But this code is also generated for us, so we don't need to worry about it at all!

Now, there are a lot of things happening behind the scenes. The good part is that they're autogenerated for us, working under the hood. Understanding how these processes work helps us appreciate the beauty of the APIs we use and makes the debugging process easier if anything goes wrong. Keep in mind that this is an advanced topic, and it's OK if it seems confusing at first. Take some time to understand these concepts, as they will make your development life easier in the long run.

Before we move on, let's take a quick look at the native iOS side of things. On the Swift side, a LocalStorageApi.swift class is generated for us. It has the same logic as the Kotlin file, so we won't be diving into it. To connect it to the actual iOS app, we need to update our AppDelegate.swift file. This is how it looked before:

ios/Runner/AppDelegate.swift

```swift
@objc class AppDelegate: FlutterAppDelegate {

    override func application(
        _ application: UIApplication,
        didFinishLaunchingWithOptions launchOptions
```

```
            : [UIApplication.LaunchOptionsKey: Any]?
            ) -> Bool {
        let controller: FlutterViewController
            = window?.rootViewController as! FlutterViewController
        let favesChannel = FlutterMethodChannel(
            name: "com.example.candy_store/faves",
            binaryMessenger: controller.binaryMessenger
        )
        favesChannel.setMethodCallHandler { [weak self] (
            call: FlutterMethodCall,
            result: @escaping FlutterResult) in guard let self =
            self else { return }
            switch call.method {
            case "getFaves":
                result(self.getFaves())
            case "addFave":
                if let args = call.arguments as? [String: Any],
                    let id = args["id"] as? String {
                    self.toggleFavorite(id, isFavorite: true)
                    result(nil)
                } else {
                    result(
                        FlutterError(
                            code: "INVALID_VALUE",
                            message: "id is null",
                            details: nil)
                    )
                }
                // ... other methods
            default:
                result(FlutterMethodNotImplemented)
            }
        }
    }
}
```

All of the MethodChannel setup, argument parsing, result returning, and so on, was done manually. When we refactor it to use the code generated by pigeon, we will get rid of all of that code:

ios/Runner/AppDelegate.swift

```
objc class AppDelegate: FlutterAppDelegate, LocalStorageApi {
    override func application(
```

```swift
        _ application: UIApplication,
        didFinishLaunchingWithOptions launchOptions
            : [UIApplication.LaunchOptionsKey: Any]?
            ) -> Bool {
        let controller: FlutterViewController
            = window?.rootViewController as! FlutterViewController
        LocalStorageApiSetup.setUp(
            binaryMessenger: controller.binaryMessenger,
            api: self
        )
        return super.application(
            application,
            didFinishLaunchingWithOptions: launchOptions
        )

    }

    func addFave(id: String) { ... }

    func removeFave(id: String) { ... }

    // and other methods from LocalStorageApi

}
```

In the same way that we did with Kotlin, we need to provide the implementation of the `LocalStorageApi` class. In our case, for simplicity, it is implemented by `AppDelegate` itself. Afterward, we need to call the `setUp` method. And that's all!

Let's summarize what we have done:

1. We described an interface in Dart using the `@HostApi` annotation.

2. Platform bindings were automatically generated for us by the `pigeon` code generator, based on this interface.

3. On the Dart side, we can now use `LocalStorageApi` as a regular Dart class.

4. On the native side, we implemented `LocalStorageApi` as a regular Kotlin or Swift interface. We then connected it by calling a single method – the `setUp` method.

5. By doing this, we have resolved all issues related to string matching and made our code type-safe, as it should be with languages such as Dart, Kotlin, and Swift. Additionally, we have eliminated the need for the `result` API, reducing the risk of bugs caused by its incorrect usage.

The `pigeon` package is a great example of how problems related to type safety can be resolved by using code generation. In the grand scheme of things, it is important to value type safety and ensure it whenever possible. During your Flutter development journey, you will encounter more situations where you will be working with dynamic code, such as network models and Firebase services. Always question yourself and search for solutions that can make your code more reliable. Before we wrap up, let's review a few more aspects and limitations of the pigeon tool.

What else you can (and can't) do with pigeon

The use case discussed in this chapter involved calling methods from the client (Flutter) to the host (native). However, there can also be a scenario where we need to call methods from the host to the client. This communication follows the same principles but in reverse. We set up the method channel handler on the Flutter side and invoke methods on the native side. The good news is that pigeon also supports this scenario. To achieve this, you just need to describe the interface using the @FlutterApi annotation instead of (or alongside) @HostApi and perform the setup steps in reverse. The bad news is that if you need to use EventChannel for streaming events, pigeon currently does not support code generation for that, so there is no alternative.

Previously, we removed the result API, which was necessary to complete asynchronous method calls. Technically, it still exists, but it is abstracted away from us by the generated code, allowing our implemented interface to imitate synchronous behavior. However, there are situations where synchronous behavior is not possible. For example, when implementing a callback that will execute in the future, we cannot simply return from the method at the end. In such cases, we would still need to use the result API. To enable it, we can annotate our @HostApi method with an @async annotation. The annotated method would look like this:

pigeon/local_storage_pigeon_api.dart

```dart
@HostApi()
abstract class LocalStorageApi {
  @async
  bool isFave(String id);
}
```

This would generate an additional argument for the isFave method on the native side. So, previously on the Kotlin side, we had a method like this:

android/app/src/main/kotlin/com/example/candy_store/MainActivity.kt

```kotlin
// #1
    override fun isFave(id: String) {
        val preferences = getSharedPreferences()
        val faves = preferences.getStringSet(
            "faves", HashSet()
```

```
    ) ?: HashSet()
    return faves.contains(id) // #2
}
```

It had a single `id` param, as you can see on line #1. After we regenerate the code with the `@async` annotation, it will look like this:

android/app/src/main/kotlin/com/example/candy_store/MainActivity.kt

```
// #1
    override fun isFave(id: String,
                        callback: (Result<Boolean>) -> Unit) {
        val preferences = getSharedPreferences()
        val faves = preferences.getStringSet(
            "faves", HashSet()
        ) ?: HashSet()
        return callback(Result.success(faves.contains(id))) // #2
    }
```

Although this reintroduces the possibility of errors related to misusing `result`, in this scenario, we explicitly generate `callback`, indicating that we are aware of the need for asynchronous behavior. This approach encourages us to be more mindful about the implementation. I believe that having the choice and explicitly marking only methods that actually require asynchronous behavior as such will contribute to code safety and readability.

The main idea of this chapter is to emphasize the importance of type safety in Flutter development. As developers, we should actively seek type safety when it's not readily available and strive to make our code resilient to unexpected behavior and potential errors. The `pigeon` code generation tool is a great solution for this purpose. However, it's important to understand that it still relies on the plain old platform channels underneath and doesn't involve any magic. Another emerging solution is the **foreign function interface** (**FFI**), which is still in the experimental stage but worth keeping an eye on. FFI allows us to call functions from other languages directly in Dart by generating code bindings. Unlike platform channels, which have strict rules on threading and data encoding, FFI enables access to native memory and allows faster communication within the same thread. Currently, as of 30.04.24, Dart and Objective-C/Swift interop via FFI (`https://dart.dev/guides/libraries/objective-c-interop`) and Dart and Java/Kotlin interop via FFI (`https://dart.dev/guides/libraries/java-interop`) are still experimental and actively being developed. Keep an eye on this for future updates.

Summary

In this chapter, we reviewed an advanced concept – working with the native layer. We observed the architecture of the Flutter application, the responsibilities of each layer, as well as how the client (Flutter) and host (native) are connected via platform channels. We implemented a **Favorites** feature on the native side via the `MethodChannel` API, which highlighted the downsides and potential problems of this default approach. Then, we learned how the `pigeon` code generation tool solves these problems by ensuring type safety and reducing the amount of boilerplate.

This is the final chapter of *Part 3*, which means that at this point, you should be well equipped with solutions and battle-proven approaches to a broad set of Flutter problems. In *Part 4*, the final part of the book, we will learn how to test the code that we have written so far, as well as what other options we have for ensuring and maintaining the code quality of our applications. We will start with *Chapter 11*, Unit Tests, Widget Tests, and Mocking Dependencies, where you will learn how to write different types of automated tests for our app.

Get this book's PDF version and more

Scan the QR code (or go to `packtpub.com/unlock`). Search for this book by name, confirm the edition, and then follow the steps on the page.

Note: Keep your invoice handy. Purchases made directly from Packt don't require an invoice.

Part 4:
Ensuring App Quality
and Stability

In the final part, you will learn how to ensure your app's quality remains high and stable as it grows and scales. We will cover the topic of automated testing, how it is done in Flutter, and how to write productive tests by leveraging the power of unit and widget testing frameworks. We will also see how static code analysis tools and Flutter DevTools help us keep our code base as clean and bug-free as possible.

This part includes the following chapters:

- *Chapter 11, Unit Tests, Widget Tests, and Mocking Dependencies*
- *Chapter 12, Static Code Analysis and Debugging Tools*

11

Unit Tests, Widget Tests, and Mocking Dependencies

Testing is a crucial part of developing robust and reliable apps. In Flutter, testing ensures that your widgets and business logic work as expected, providing a seamless experience for users. This chapter will introduce you to the fundamental aspects of testing in Flutter, focusing on unit tests, widget tests, and mocking dependencies.

In this chapter, we'll guide you through writing unit tests to verify the core logic of your Flutter app. You'll learn how to create and execute widget tests to ensure your UI components render correctly and interact as expected. Additionally, we'll cover how to mock dependencies using `mockito`, which allows you to isolate and test different parts of your app without having to rely on actual implementations.

By the end of this chapter, you will have a solid understanding of the various testing methodologies in Flutter and how to apply them to your projects. You will be able to write tests that improve your app's quality and maintainability with confidence.

In this chapter, we're going to cover the following main topics:

- Getting started with unit testing
- Widget testing fundamentals
- Mocking dependencies for effective testing

Technical requirements

To get started with this chapter, you'll need to check out the CH11 initial branch. It contains a simple Flutter app that's been created by running a `flutter create` command and adding some simple model, repository, service, and widget code to make things easier for us: https://github.com/PacktPublishing/Flutter-Design-Patterns-and-Best-Practices/tree/master/CH11/initial/candy_store.

You can find the final working code here: https://github.com/PacktPublishing/Flutter-Design-Patterns-and-Best-Practices/tree/master/CH11/final/candy_store.

You can just create a test folder at the same level as the lib folder in the file hierarchy and start following the instructions provided.

Getting started with unit testing

In this section, we'll explore the basics of **unit testing** in Flutter. Unit tests are essential for verifying that individual pieces of your app's code work as intended. By focusing on the smallest parts of your app, such as functions and classes, unit tests help ensure that each component behaves correctly in isolation.

Unit testing is particularly useful because it allows developers to catch bugs early in the development process, improving the reliability and maintainability of the code. In this section, you'll learn how to write unit tests for your app's core logic using Flutter's testing framework.

By the end of this section, you will have a solid understanding of how to create and run unit tests in Flutter. You'll be equipped with the skills to ensure that your app's core functionality is robust and free of defects.

To effectively test our cart logic, we need a way to simulate backend interactions without relying on an actual backend. This is where a fake repository comes in.

Introducing a fake repository

A fake repository is a mock implementation that mimics the behavior of a real repository. It allows us to test our business logic in isolation, without the need for an actual backend service. Check out the CartRepository interface to understand the fake repository better. The CartRepository interface defines the methods that any created CartRepository should implement. It ensures that any concrete class, such as FakeCartRepository, adheres to this contract. So, considering that, let's create the FakeCartRepository class in a new file named fake_cart_repository.dart in the test directory:

```
class FakeCartRepository implements CartRepository {
  final _cartInfoController = StreamController<CartInfo>.broadcast();
  final Map<String, CartListItem> _items = {};
  double _totalPrice = 0;

  @override
  Stream<CartInfo> get cartInfoStream => _cartInfoController.stream;

  @override
  Future<CartInfo> get cartInfoFuture async {
    return CartInfo(
```

```
    items: Map.from(_items),
    totalPrice: _totalPrice,
    totalItems: _items.length,
  );
}
```

Here, we're creating `StreamController` to manage a stream of `CartInfo` objects. This will allow us to emit updates about the cart's states for `FakeCartRepository`. Then, we can create `_items` to store items in the cart, and we can use their `itemId` values as the key and `CartListItem` as the value. As we're adding the items, we should ensure we have a variable to keep track of the total price of items in the cart. This is where `totalPrice` comes in.

Now, we can start explaining the `cartInfoStream` getter, which is responsible for returning the stream of `CartInfo` updates. It allows listeners (such as `CartBloc`) to be notified whenever the cart's state changes. The `cartInfoFuture` getter returns a `Future` object that resolves to the current `CartInfo` object, encapsulating the cart's state. Now, let's add the other necessary functions – that is, `addToCart`, `removeFromCart`, and, of course, the disposal function:

```
@override
Future<void> addToCart(ProductListItem item) async {
  if (_items.containsKey(item.id)) {
    final currentItem = _items[item.id]!;
    _items[item.id] = CartListItem(
      product: currentItem.product,
      quantity: currentItem.quantity + 1,
    );
  } else {
    _items[item.id] = CartListItem(product: item, quantity: 1);
  }
  _totalPrice += item.price;
  _cartInfoController.add(await cartInfoFuture);
}


@override
Future<void> removeFromCart(CartListItem item) async {
  if (_items.containsKey(item.product.id)) {
    final currentItem = _items[item.product.id]!;
    _totalPrice -= item.product.price;
    if (currentItem.quantity > 1) {
      _items[item.product.id] = CartListItem(
        product: currentItem.product,
        quantity: currentItem.quantity - 1,
      );
    } else {
```

```
        _items. Remove(item.product.id);
      }
      _cartInfoController.add(await cartInfoFuture);
    }
  }

  void dispose() {
    _cartInfoController.close();
  }
}
```

First, `addToCart` adds the item to the cart. If the item already exists, it increases the quantity. Otherwise, it adds a new `CartListItem`. After updating the cart, it emits the updated `CartInfo` via the stream. Meanwhile, `removeFromCart` is performing similar functionality by decreasing the quantity and removing the item entirely. Finally, our `dispose()` method closes the stream controller when it is no longer needed, preventing memory leaks.

Since we have `FakeCartRepository`, we can also create our test data so that we can write our unit tests.

Creating test data

This is the simplest step of our tests. Here, we'll create one `List<Product>` and one `ProductListItem` that we can use in our tests and know what to expect in terms of the results. We can just copy any example we have and adjust the data as needed. The following code shows our `TestData` class, which we'll use in the following examples. Please create a file named `test_data.dart` under the test directory:

```
import 'package:candy_store/product/domain/model/product.dart';
import 'package:candy_store/product/domain/model/product_list_item.
dart';

class TestData {
  static const List<Product> testProducts = [
    Product(
      id: '2',
      name: 'Test Donut',
      description: 'A soft and sweet test donut, glazed or filled
with your favorite flavors',
      price: 4,
      imageUrl: 'resources/images/donut.webp',
      sku: '231fd',
```

```
        stock: 5,
      ),
    ];

    static final testProductListItem = ProductListItem(
      id: '1',
      name: 'Test Bean',
      description: 'Colorful and chewy test beans in a variety of
fruity flavors',
      price: 2,
      imageUrl: 'resources/images/jellybean.webp',
    );
}
```

All objects are created as defined in the classes in previous chapters. You can *Cmd + Click* your code and forward it to the related file in your editor. With this `TestData`, we have a consistent set of objects to use across our tests. This ensures that we always know the expected results when we're running our tests. For instance, if we add `testProductListItem` to the cart, we can confidently check that the item's `name` is `Test Bean`, and its `price` is 2. This consistency helps us verify that the tests are behaving correctly and that the app's logic is working as intended. At this point, we can start writing our unit tests.

Writing unit tests

We need to write unit tests to ensure that our methods work correctly. Unit tests will help us verify that our adding to cart, removing from the cart, and initializing methods behave as expected.

Understanding the testing structure

Before diving into the actual tests, let's understand the structure and functions we'll use:

- `main()`: The entry point of our test file
- `group()`: Organizes tests into a group, allowing us to categorize them
- `test()`: Defines a single test case
- `expect()`: Defines the expected outcome of a test

Now that you understand the testing structure, we can start writing our test.

First, we'll create a new file named `unit_test.dart` in the test directory. Then, we need to call our fake repository, use it for our `bloc` function, call our product, and add it as our `cartItem` to start working on it. After, we can populate our item, load it, and remove it from our cart. Being able to successfully remove an item from the cart will be our goal and the focus of our unit test. Let's check out the code together; we'll explain it step by step after:

```
test('Remove item from cart', () async {
  final fakeCartRepository = FakeCartRepository();
  final cartBloc = CartBloc(cartRepository: fakeCartRepository);
  final product = TestData.testProductListItem;
  final cartItem = CartListItem(product: product, quantity: 1);

  await fakeCartRepository.addToCart(product);
  cartBloc.add(const Load());

  await Future.delayed(const Duration(milliseconds: 200));

  final states = <CartState>[];
  final subscription = cartBloc.stream.listen(states.add);

  cartBloc.add(RemoveItem(cartItem));

  await Future.delayed(const Duration(milliseconds: 500));
  await subscription.cancel();

  expect(states.length, greaterThanOrEqualTo(3));
  expect(states[0].loadingResult.isInProgress, isTrue);
  expect(states[1].loadingResult.isInProgress, isFalse);
  expect(states[2].items.length, equals(0));
  expect(states[2].totalPrice, equals(0));
});
```

In the preceding code, you can see that `FakeCartRepository` simulates the cart repository. An instance of `CartBloc` is created using the fake repository. Then, a test product is retrieved from `TestData`. After, `CartListItem` is created using the test product with a quantity of 1. The `addToCart` line adds the test product to the cart using the fake repository. This ensures that the cart initially contains one item before the test proceeds.

At this point, we can go for the `Cart` part; the `Load` event is added to `cartBloc` to load the initial state of the cart. Here, `Future.delayed` provides a short delay to allow the state change to be processed. We're using a list named `states` to store the cart states and the test subscribes to the `cartBloc` stream, adding each emitted state to the `states` list. We'll use this list while we're checking the results. Now, we can continue to the action part.

The `RemoveItem` event is added to `cartBloc`, which causes the specified item to be removed from the cart. Again, we're using `Future.delayed` to provide a short delay, after which we can cancel the stream subscription since we don't need any further changes to be tracked.

To read the expectation and result part, the test verifies that at least three states were emitted. We'll check each state one by one. The first test checks that the first state has a loading result that is in progress (isInProgress is true). The second test confirms that the second state has a loading result that is not in progress (isInProgress is false). The final test ensures that the third state (after the item is removed) has no items in the cart and the total price is zero.

At this point, the code looks fine, so it's time to run the unit tests and see the result. To run your unit tests, open a Terminal and navigate to your Flutter project directory. Then, run the following command:

```
flutter test
```

This command runs all the tests in your project and provides a summary of the results. Our expected result should say **All tests passed!**

By writing these unit tests, we ensured that the RemoveItem functionality in CartBloc works correctly. This foundational step ensures that our basic data structures behave as expected, providing a solid base for further development.

Next, we'll move on to widget testing, where we'll verify that our UI components render correctly and interact as intended. This transition from unit testing to widget testing will ensure that both your app's logic and its UI are robust and reliable.

Widget testing fundamentals

In this section, we'll explore the fundamentals of widget testing in Flutter. Widget tests, also known as component tests, verify that the widgets in your app behave as expected. These tests are more comprehensive than unit tests as they cover the UI and user interactions.

Widget testing ensures that your app's UI components render correctly and interact with each other as intended. By simulating user interactions, widget tests help catch UI bugs and ensure a seamless user experience.

By the end of this section, you'll be able to create and run widget tests to verify the behavior of your Flutter app's UI components.

Implementing a golden file test

Golden testing is a form of visual regression testing. In Flutter, golden tests compare the rendered UI of a widget to a reference image that we call a "golden" image. This helps ensure that the UI doesn't change unexpectedly. If the widget's appearance changes, the test fails, indicating a visual regression. To specify the difference between widget testing and golden testing, widget testing checks the functionality and appearance of individual widgets in isolation while golden testing compares the visual output of widgets to a reference image to detect unintended visual changes. Our example sets up a widget for testing, initializes the test environment, and captures a golden image for comparison.

To perform golden tests, we'll need the `golden_toolkit` package. We need to add this to our `pubspec.yaml` file under our development dependencies:

```
golden_toolkit: ^0.15.0
```

Now, we can start.

Creating the CandyShopGoldenTestWidgetsPage widget

In this section, we'll create a widget page that includes the widgets we want to test visually. This page will be used as a reference for our golden tests. You can find our `ProductListItemView` widget under `lib/product/presentation/widget/product_list_item_view.dart`. Additionally, our `CartButton` widget can be found under `lib/cart/presentation/widget/cart_button.dart`. We'll also add some text to enrich our golden test:

```
Scaffold(
  appBar: AppBar(
    title: const Text('Design System'),
  ),
  body: SingleChildScrollView(
    padding: const EdgeInsets.all(16.0),
    child: Column(
      crossAxisAlignment: CrossAxisAlignment.start,
      children: <Widget>[
        ProductListItemView(
          item: TestData.testProductListItem,
          showPlaceholderForImage: true,
        ),
        const CartButton(
          count: 2,
        ),
        Text('bodyLarge', style: Theme.of(context).textTheme.
bodyLarge),
        Text('bodyMedium', style: Theme.of(context).textTheme.
bodyMedium),
        Text('labelMedium', style: Theme.of(context).textTheme.
labelMedium),
      ],
    ),
  ),
);
```

This page includes `ProductListItemView`, `CartButton`, and some text styles. It will serve as our test case for visual inspection. Now that we have our `CandyShopGoldenTestWidgetsPage` widget, let's write some widget tests to verify its behavior.

Writing widget tests

These tests will check if the tasks are displayed correctly, as well as if the toggle and delete functionalities work as expected.

Before diving into the actual tests, let's understand the structure and functions we'll be using:

- `main()`: The entry point of our test file
- `testWidgets()`: Defines a single widget test case
- `pumpWidget()`: Renders the widget under test
- `find()`: Locates widgets in the widget tree

Now that you understand the structure of widget tests, we can start writing one.

Create a new file named `golden_widget_test.dart` in the test directory. Then, follow these steps:

1. First, we'll pump our widgets and try to match them with the golden file. So, let's start with our first test:

    ```
    import 'package:flutter/material.dart';
    import 'package:flutter_test/flutter_test.dart';
    import 'candy_shop_golden_test_widgets_page.dart';

    void main() {
      testWidgets('Golden test', (WidgetTester tester) async {
        TestWidgetsFlutterBinding.ensureInitialized();
        await tester.pumpWidget(const MaterialApp(
          home: CandyShopGoldenTestWidgetsPage(),
        ));
        await tester.pumpAndSettle();
        await expectLater(
          find.byType(CandyShopGoldenTestWidgetsPage),
          matchesGoldenFile('goldens/candy_shop_widgets.png'),
        );
      });
    }
    ```

    When you run this test, you'll notice that it's failing. This is because we need to create the relevant golden files.

2. Now, we can proceed to the next step by creating the golden file:

```
flutter test --update-goldens test/golden_widget_test.dart
```

As you can see, after you run this, the golden files will be created in the specified location. You can find your file there in PNG format. Right now, you will see that the text isn't rendered. It's possible to load fonts, but to keep our content focused, you can find the commented-out version in our test file in this book's GitHub repository and proceed from there. Loading fonts will bring other issues, advantages, and disadvantages. So, we'll set this to one side for now.

3. Once you've generated your file, you can run the test again. You will see that it passes.

To see how golden files work, we'll make the test fail intentionally. To do so, change anything in our `CandyShopGoldenTestWidgetsPage`. For example, let's make the `CartButton` widget's `count` value 1. If you run the golden test, you will see that the test framework notifies you that the golden test has failed. The output typically includes the location of the expected golden image and the actual captured image. You can find all related images in the `failures` folder, which will be generated alongside the output. You can check and either fix the code changes you've made, or you may make the change intentionally. If it's intentional, then you need to update the golden image to reflect its new appearance.

4. Run the following command once more to generate your golden file:

```
flutter test --update-goldens test/golden_widget_test.dart
```

Once you've updated your test, the expected result should say **All tests passed!** This result will verify the look of our golden test widget. This step ensures the UI components render correctly, changes are made intentionally, and that the visual integrity of your app is maintained. Now, you can even see the changes without running your main code in a simulator and catch the unintended changes even earlier.

Next, we'll move on to mocking dependencies, where we'll learn how to simulate external services and repositories to isolate and test different parts of our app. This transition from widget testing to mocking dependencies will help ensure that your app's components work correctly in isolation.

Integration testing fundamentals

These tests verify the behavior of a complete app or a significant part of it, often involving multiple widgets and interactions. They ensure that the whole app works together as expected.

You can use the `integration_test` package to run the app on a device or emulator, testing the full app's interaction with real services and user flows. Also, you can add Firebase Test Lab so that you have support for multiple devices.

In this section, we'll work on an improved version of widget testing that will interact between various components and logic to help us avoid having to use integration tests that will be run on a real device.

On top of that, it will effectively validate the interactions between various parts of the app, making them closer to integration tests as defined by the Flutter documentation.

Creating a fake repository

Recall that we created `FakeCartRepository` at the beginning of this chapter. At this point, we also need to create a fake product repository since our tests need to validate more interactions between components. So, create a `fake_product_repository.dart` file in our current test folder and add the following code:

```
class FakeProductRepository implements ProductRepository {
  @override
  Future<List<Product>> fetchProducts() async {
    return TestData.testProducts;
  }

  @override
  Future<List<Product>> searchProducts(String query) async {
    return TestData.testProducts.where((product) => product.name.
      contains(query)).toList();
  }
}
```

The `FakeProductRepository` class implements the `ProductRepository` interface, providing hardcoded responses for fetching and searching products. This allows us to control the data that's returned during tests. To ensure we're always using the same values, we're using `TestData`, which we created previously.

Displaying an item in the cart test

Since we also have the product repository now, we can verify the action of showing the candy and its total pricing in `CartPage` after the user adds a product to the cart.

Let's create our `integration_test.dart` file under the test folder. Add the following code to it:

```
testWidgets('should display an item in the cart', (WidgetTester
tester) async {
    final fakeCartRepository = FakeCartRepository();
    final cartBloc = CartBloc(cartRepository: fakeCartRepository);

    final product = TestData.testProductListItem;
    await fakeCartRepository.addToCart(product);

    await tester.pumpWidget(
      MaterialApp(
```

```
        home: BlocProvider<CartBloc>(
          create: (_) => cartBloc,
          child: const CartPage(),
        ),
      ),
    );
    await tester.pump();

    expect(find.text('Test Bean'), findsOneWidget);
    expect(find.text('Total:'), findsOneWidget);
    expect(find.text('2.0 €'), findsOneWidget);
  });
```

Now, as you go through the code, you'll see steps similar to the ones we passed in previous sections. Here, we're creating an instance of `FakeCartRepository` and `CartBloc`, adding the product to the repository, pumping it to ensure the widget tree is built, and verifying that the product's name and price are displayed correctly in the cart.

Creating the widget test with interaction

Widget tests don't run like integration tests in a real device because they don't need to. However, there are some ways to make them interact with our app and perform the related taps and calculations. Let's check out our more advanced widget test, which contains interactions:

```
testWidgets('Add one product to cart in main page.', (WidgetTester
tester) async {
    final fakeCartRepository = FakeCartRepository();
    final fakeProductRepository = FakeProductRepository();
    final cartBloc = CartBloc(cartRepository: fakeCartRepository);

    await tester.pumpWidget(
      MultiRepositoryProvider(
        providers: [
          RepositoryProvider<CartRepository>(
            create: (_) => fakeCartRepository,
          ),
          RepositoryProvider<ProductRepository>(
            create: (_) => fakeProductRepository,
          ),
        ],
        child: MultiBlocProvider(
          providers: [
            BlocProvider<CartBloc>(
              create: (_) => cartBloc,
```

```
          ),
        ],
        child: MaterialApp(
          home: MainPage.withBloc(),
        ),
      ),
    ),
  );

  await tester.pumpAndSettle();
  expect(find.widgetWithText(CartButton, '0'), findsOneWidget);
  await tester.tap(find.byIcon(Icons.add));
  await tester.pumpAndSettle();
  expect(find.widgetWithText(CartButton, '1'), findsOneWidget);
});
```

Once again, we're setting up mocks and blocs to simulate the app's state and interactions without relying on actual backend services. You may notice an unfamiliar section of code that contains pumpWidget and MultiRepositoryProvider. This setup ensures that the widget tree can access the necessary dependencies (repositories and blocs) for the test. This is exactly what we have in the app.

You can also see that we have our simulation of a user tapping the add button (represented by an Icons.add icon) to add a product to the cart. It triggers our app's business logic to update the cart state. We're also checking that the state is updated correctly in response to user action, reflecting the addition of one product.

While connecting the concepts, we can benefit from the aid of mockito, which helps us to simulate the behavior of real objects in a controlled way. It allows us to isolate and test specific parts of our app without having to rely on actual implementations. In previous steps, we created our own behaviors in files. Now, we'll try this out the mockito way!

Mocking dependencies for effective testing

In this section, we'll explore how to mock dependencies in Flutter using **mockito**. **Mocking** is a technique that's used to simulate the behavior of real objects in a controlled way. As mentioned previously, this allows us to isolate and test specific parts of our app without having to rely on actual implementations. Mocking is especially useful for testing components that depend on external services or complex interactions.

By the end of this section, you'll be able to use mockito to mock services and repositories, enabling you to test your app's components in isolation. This will improve the reliability and maintainability of your tests.

Please check out the mockito_example folder in this book's GitHub repository. There, you will find the mockito way of implementing the dependencies.

Understanding mocking and mockito

We can use mocking to isolate and test specific components, simulate different scenarios and edge cases, and improve test reliability and speed by avoiding dependency on external services.

mockito is a Dart package that's used for creating mock objects. It allows us to specify the behavior of mock objects and verify interactions with them.

Setting up mockito and mocking dependencies

Before we start mocking, we need to set up our project so that it can use mockito and build_runner:

1. Open the pubspec.yaml file and add the following dependencies under dev_dependencies:

    ```
    dev_dependencies:
      flutter_test:
        sdk: flutter
      mockito: ^5.0.0
      build_runner: ^2.0.0
    ```

    Don't forget to run flutter pub get to install these dependencies after saving the file. We need build_runner to create the mock classes that you'll see in the next step.

2. Create a file named cart_repository.dart in the test/mockito_example directory and add the following content:

    ```
    import 'package:mockito/annotations.dart';
    import 'package:candy_store/cart/domain/repository/cart_
    repository.dart';

    @GenerateMocks([CartRepository])
    void main() {}
    ```

3. To generate the mock classes that we specified with @GenerateMocks, run the following command:

    ```
    flutter pub run build_runner build
    ```

This command generates a file named cart_repository.mocks.dart that contains the MockCartRepository class.

Writing the unit test with mockito

In this section, we'll write tests for `CartRepository` using the mock service.

You can find the code for this section in this book's GitHub repository; it will allow you to follow and understand this process more easily. First, we have the variable declarations:

test/mockito_example/mockito_unit_test.dart

```
late MockCartRepository mockCartRepository;
late CartBloc cartBloc;
late StreamController<CartInfo> cartInfoController;
```

These variables will be used in the tests. Here, `mockCartRepository` is a mock of `CartRepository`, `cartBloc` is the bloc we are testing, and `cartInfoController` is a `StreamController` class that's used to manage the stream of `CartInfo` updates. Next, we'll follow up with the setup function:

test/mockito_example/mockito_unit_test.dart

```
setUp(() {
  mockCartRepository = MockCartRepository();
  cartInfoController = StreamController<CartInfo>.broadcast();
  when(mockCartRepository.cartInfoStream).thenAnswer((_) =>
    cartInfoController.stream);
  cartBloc = CartBloc(cartRepository: mockCartRepository);
});
```

This function runs before each test. It initializes the mock repository and `StreamController`, sets up the mock behavior for `cartInfoStream`, and creates an instance of `CartBloc` using the mocked repository:

test/mockito_example/mockito_unit_test.dart

```
tearDown(() {
  cartInfoController.close();
});
```

This function runs after each test. It ensures that `StreamController` is closed properly so that resources can be cleaned up and memory leaks can be avoided:

test/mockito_example/mockito_unit_test.dart

```dart
test('Initial state is correct', () {
  expect(
    cartBloc.state,
    const CartState(
      items: {},
      totalPrice: 0,
      totalItems: 0,
      loadingResult: DelayedResult.idle(),
    ),
  );
});
```

This test checks that the initial state of `CartBloc` is correct. It ensures that when `CartBloc` is first created, its state has no items, a total price of 0, a total item count of 0, and an `idle` loading result:

test/mockito_example/mockito_unit_test.dart

```dart
test('Remove item from cart', () async {
  final product = TestData.testProductListItem;
  final cartItem = CartListItem(product: product, quantity: 1);
  final initialCartInfo = CartInfo(
    items: {product.id: cartItem},
    totalPrice: product.price.toDouble(),
    totalItems: 1,
  );
  final emptyCartInfo = CartInfo(
    items: {},
    totalPrice: 0,
    totalItems: 0,
  );
```

We can begin the test setup for removing an item from the cart. It initializes the product and `cartItem`, which represents a product and its instance in the cart. It also sets up `initialCartInfo` to represent the cart's state before the removal and `emptyCartInfo` to represent the cart's state after the removal:

test/mockito_example/mockito_unit_test.dart

```
when(mockCartRepository.cartInfoFuture).thenAnswer((_) async =>
initialCartInfo);
when(mockCartRepository.removeFromCart(cartItem)).thenAnswer((_) async
{
  await Future.delayed(const Duration(milliseconds: 50)); // Simulate
network delay
  when(mockCartRepository.cartInfoFuture).thenAnswer((_) async =>
    emptyCartInfo);
  cartInfoController.add(emptyCartInfo);
});
```

Now, we must set up mock behavior for `CartRepository`. Initially, `cartInfoFuture` returns `initialCartInfo`. When `removeFromCart` is called, it simulates a network delay, then updates `cartInfoFuture` so that it returns `emptyCartInfo` and adds `emptyCartInfo` to the `cartInfoController` stream:

test/mockito_example/mockito_unit_test.dart

```
await mockCartRepository.addToCart(product);
cartBloc.add(const Load());
await Future.delayed(const Duration(milliseconds: 200));
```

Next, we must pre-populate the cart with the product to set up the initial state for the test. This involves calling `addToCart` on the repository and dispatching a `Load` event to `CartBloc`:

test/mockito_example/mockito_unit_test.dart

```
final states = <CartState>[];
final subscription = cartBloc.stream.listen(states.add);
```

We can also add a list to collect state changes and start listening to the `CartBloc` stream, adding each state to the list as it changes:

test/mockito_example/mockito_unit_test.dart

```
cartBloc.add(RemoveItem(cartItem));
await Future.delayed(const Duration(seconds: 1));
await subscription.cancel();
```

With this snippet, we're dispatching a `RemoveItem` event to `CartBloc` and waiting for the event to be processed. After a delay, cancel the subscription to stop listening to state changes:

test/mockito_example/mockito_unit_test.dart

```
expect(states.length, greaterThanOrEqualTo(3));
expect(states[0].loadingResult.isInProgress, isTrue);
expect(states[1].loadingResult.isInProgress, isFalse);
expect(states[2].items.length, equals(0));
expect(states[2].totalPrice, equals(0));
```

As a result, we check every step, ensure there are at least three states, check the loading states, and verify that the final state reflects that the item has been removed from the cart.

This test setup and execution ensures that `CartBloc` correctly handles the removal of an item from the cart. By using `mockito` to mock `CartRepository`, the test isolates the behavior of `CartBloc` and verifies its interactions with the repository, ensuring that the bloc emits the correct states throughout the process.

In this section, we learned how to use `mockito` to mock dependencies in Flutter.

Summary

In this chapter, you learned how to conduct thorough testing in Flutter using unit tests, widget tests, and mocking dependencies. We started by exploring unit tests, which verify the functionality of individual pieces of code, such as functions and classes. This included understanding the basic structure of unit tests and how to run them. Then, we moved on to widget tests, which ensure that UI components render correctly and interact as expected. There, you learned how to write widget tests to verify the behavior of your Flutter app's UI components. Finally, we delved into mocking dependencies using `mockito`, which allows you to simulate external services and repositories, allowing you to test your app's components in isolation without having to rely on actual implementations.

Understanding unit testing in Flutter provides a solid foundation for ensuring that your app's core logic works as intended. Learning how to transition from testing individual units to testing UI components through widget tests equips you with the skills to verify both the functionality and user experience of your app. Finally, mastering the use of `mockito` for mocking dependencies enhances your ability to create reliable and maintainable tests, ensuring that your app's components work correctly in isolation. Collectively, these skills enable you to build robust, high-quality Flutter apps, whether you're focusing on core logic, UIs, or complex interactions with external services.

In the next chapter, we'll talk about code quality, how to maintain it, and debugging practices and tools.

12

Static Code Analysis and Debugging Tools

Why does code quality matter? People write whole books on this topic just to answer this question. Ultimately, one reason should be enough – we write code for people, not for computers. Computers don't "care" about the quality of your code, the patterns that you use, or what you name your variables – in the end, all code gets compiled into a sequence of binary signals, which the computer just executes without "caring". However, the developers who will maintain this code care. Developers who will continue working with this code by adding new features and scaling the system to support the growing user flow – they care. Those other developers could actually be you in the future. There are so many perspectives from which code quality is a crucial matter:

- Adhering to team standards for code quality is a prerequisite for productive work. Otherwise, you can find yourself spending more time fighting and fixing rather than developing functionality.

- Having and following those standards makes the job easier not only for developers already working on the project but also for onboarding new developers, especially when there are automated tools integrated into the process.

- Messy code bases introduce not only difficulties for developers but also bugs that frustrate users. The harder it is to navigate the codebase, the more time it takes to fix even simple bugs, negatively impacting the end users and the business.

That's why we care about such things as code base consistency, maintainability, scalability, durability, and testability. That's why we wrote a whole book on how to follow design patterns and best practices.

In this last chapter, we will review how we can ensure that the quality of our code base is consistently maintained, which tools we can use, and what our options are when it comes to debugging the existing issues. In this chapter, we're going to cover the following main topics:

- Following coding conventions
- Ensuring consistency and preventing bugs with static code analysis
- Exploring debugging practices and tools

Technical requirements

In order to proceed with this chapter, you will need the following:

- Code from the previous chapter, which can be found here: `https://github.com/PacktPublishing/Flutter-Design-Patterns-and-Best-Practices/tree/master/CH11/final/candy_store`.

- Access to all of the code required for this chapter, which you can find here:

  - The code for the start of the chapter can be found at `https://github.com/PacktPublishing/Flutter-Design-Patterns-and-Best-Practices/tree/master/CH12/initial/candy_store`.

  - The code for the end of the chapter is available at `https://github.com/PacktPublishing/Flutter-Design-Patterns-and-Best-Practices/tree/master/CH12/final/candy_store`. You can review the step-by-step refactoring in the commit history of this branch.

Following coding conventions

Code consistency starts early on, with code formatting. Properly formatted code is easier to read, navigate, and update. While the specifics of what constitutes "properly formatted code" may vary between languages, teams, or individuals, Dart and Flutter have some widely accepted conventions.

For example, some of the Dart and Flutter conventions include the following:

- Limit line length to 80 characters
- Use snake case for file names, but camel case for class names (for example, `my_app.dart`, not `MyApp.dart`)
- Organize class code in the following order: constructors, factories, fields, methods, and so on

Obviously, everything that applies to Dart also applies to Flutter. However, there is one specific rule that is widely used in the Flutter context: the **trailing comma**. Due to the nature of Flutter code, where widget trees are often built with multiple levels of nesting, it is considered good practice to use a trailing comma. This means always putting a comma at the end of a longer list of parameters, allowing for proper formatting. To better visualize this, let's compare how a widget tree would look with and without trailing commas.

Without trailing commas, it would look like this:

```
Widget build(BuildContext context) {
return Container(width: 100, height: 100, color: Colors.red, padding:
EdgeInsets.all(16));
}
```

With trailing commas, it instead looks like this:

```
Widget build(BuildContext context) {
    return Container(
      width: 100,
      height: 100,
      color: Colors.red,
      padding: EdgeInsets.all(16),
    );
  }
```

While it may not make a logical difference for the machine or the compiler, it does make a significant difference for a human working with the code. However, how do we actually format the code? We certainly don't have to do it manually, right?

Using the dart format command

The Dart SDK provides a great environment and an array of useful tools. One such tool is the `dart format` command in the Dart **Command Line Interface** (**CLI**), which automatically formats code according to the official Dart conventions. You can use it in different ways, depending on your context:

- In the Dart CLI, use the `dart format` command.

- Depending on your IDE, you can use an extension or plugin that is bound to a specific key combination, such as *Cmd* + *Opt* + *L* on macOS in Android Studio. You can also enable the **format code on save** option, which will automatically format the current file when you save it.

Developing a habit of code formatting will do wonders to improve your productivity in the future. Find the most automated way to format your code in your environment and use it consistently. Code formatting is important, but code style conventions go beyond that. Let's explore how we can further enhance code consistency by following the **Effective Dart** suggestions.

Following the Effective Dart conventions

Writing a significant amount of Dart code and receiving feedback from users and contributors has led the Dart team to create a comprehensive and easy-to-follow guide called *Effective Dart*. You can find this guide here: `https://dart.dev/effective-dart` (as of 2024). It covers various aspects such as style, documentation, usage, and design of your Dart code. Each section provides guidelines categorized as **DO, DON'T, PREFER, AVOID,** and **CONSIDER,** indicating the level of requirement for each guideline. While there are no strict requirements to follow these guidelines, it is highly recommended to maintain consistency, efficiency, and adherence to industry standards in your code base. When you are unsure about how to approach something in Dart, you can refer to this guide as the source of truth.

As an example, let's examine this guideline: `https://dart.dev/effective-dart/usage#dont-explicitly-initialize-variables-to-null`.

The guideline states: **DON'T explicitly initialize variables to** `null`, with a subheading of **Linter rule: avoid_init_to_null**. The URL `https://dart.dev/tools/linter-rules/avoid_init_to_null` is included in the form of a hyperlink.

If a variable has a non-nullable type, Dart reports a compile error if you try to use it before it has been definitely initialized. If the variable is nullable, then it is implicitly initialized to `null` for you. There's no concept of uninitialized memory in Dart and no need to explicitly initialize a variable to `null` to be "safe."

```
// Good:
Item? bestDeal(List<Item> cart) {
  Item? bestItem;
  // ... code that uses bestItem
}

// Bad:
Item? bestDeal(List<Item> cart) {
  Item? bestItem = null;
  // ... code that uses bestItem
}
```

The guideline is comprehensive, containing a title, a description explaining the purpose of the rule, and both good and bad usage examples. This makes it convenient and easy to understand and follow. However, memorizing all of the rules can be impractical, especially when working in teams. Thankfully, there is a solution to automate this process: the linter rule mentioned at the top of the guide. Let's explore how **static code analysis** and linters can help us ensure code consistency without the need to memorize every convention.

Ensuring consistency and preventing bugs with static code analysis

Static code analysis is a well-known method for evaluating code to identify code violations and patterns that can lead to bugs. This analysis is performed without actually running the code, hence the term "static." It is a powerful tool that helps improve code quality during the compilation stage. It is sometimes referred to as a linter because it reports issues based on a set of rules known as **lint rules**. IDEs often highlight, or **lint**, problematic code lines. So, in this context, when we use the terms analyzer or linter, we are referring to the same thing, even though there is a technical difference between the two.

The Dart SDK includes another handy tool called the Dart analyzer, which performs static code analysis. Like `dart format`, you can use it in various environments. It is integrated into many popular IDEs and can also be used as a CLI command using `dart analyze`, for example, on a **Continuous Integration (CI)** system.

It is important to distinguish between the two types of warnings reported by the analyzer:

- **Compilation or syntax issues**: There are cases when the code you've written is technically incorrect, such as a wrong type, an unknown keyword, or a missing import, and that code won't even compile in the first place. This can be fixed by correcting the syntax.

- **Potential bugs or code style inconsistencies**: If your code won't even compile in the previous case because of an error, then these types of issues are potentially more dangerous. What can be perfectly valid and compilable Dart code can be an obvious bug in the context of Flutter – for example, forgetting to dispose of a controller in a `StatefulWidget` can lead to potential memory leaks. Configuring the analyzer to catch such issues can dramatically improve the stability of the code base and pace of development by avoiding issues in the first place.

By default, the Dart analyzer uses a list of the most crucial lints, such as those that ensure Dart's sound null safety. This is beneficial because it allows for flexible configuration based on team agreements and requirements. Let's explore how you can customize your list of lints.

Customizing lint rule list in analysis_options.yaml

All Flutter projects created with `flutter create`, starting from Flutter SDK version 2.3.0, have a default set of lint rules enabled. These rules are provided by the `flutter_lints` package (https://pub.dev/packages/flutter_lints), which builds on top of the Dart `lints` package (https://pub.dev/packages/lints). You can verify this by checking the `dev_dependencies` section of your project's `pubspec.yaml` file.

Having these default lint rules is a great starting point, as they contribute significantly to code consistency and safety. The `flutter_lints` package contains a carefully selected list of rules recommended by the Flutter and Dart teams. However, it's important to note that only rules that make sense for the majority of users are included in this list. The package is periodically updated with new lint rules based on user feedback, usually once a year.

While the default lint rules are a good foundation, you may want to adopt stricter and more opinionated lint rules, especially considering that many more rules are available in the Dart SDK (`https://dart.dev/tools/linter-rules`). Let's add some new rules to our `candy_store` app.

For example, let's say that we agree with our team that whenever we use hard-coded strings, we want them to be in double quotes. In Dart, you can configure a rule regarding quotes: you can force them to be single quotes or double quotes, or you can choose not to force anything at all. Up to this point in our code base, we have been using single quotes, so let's enable double quotes and see where it gets us.

In order to customize the analyzer in your Flutter project, you need to create the `analysis_options.yaml` file in your root directory, alongside the `pubspec.yaml` file. In that file, you can describe the desired behavior by making use of `include`, `analyzer`, `linter`, and other entries. We will see some of them in action in the next section. If for any reason you need to override that behavior in your sub-packages, you can do that by introducing another `analysis_options.yaml` file on the sub-package level too. It will then take precedence over the root one. Now, let's continue with our implementation:

1. First, open the `analysis_options.yaml` file, which is located alongside `pubspec.yaml`. You should see this content:

    ```
    include: package:flutter_lints/flutter.yaml

    linter:
      rules:
    ```

    We can see that the lints from the `flutter_lints` package are included, but other than that the rules section is empty. Keep in mind that we're working with the YAML format, so in this case, tabs and spaces matter.

2. Under the rules section, we want to add a lint with the `prefer_double_quotes` title and set its value to `true`. It should now look like this:

    ```
    include: package:flutter_lints/flutter.yaml

    linter:
      rules:
        prefer_double_quotes: true
    ```

3. Then run the analyzer (which depends on your IDE or CLI). In the analyzer output, you should see a long list of reported lint warnings that say something like the following:

```
info: Unnecessary use of single quotes. (prefer_double_quotes at
[candy_store] lib/cart_bloc.dart:1)
```

There are a few things to unpack here:

- The `info` prefix means that the severity level of this issue isn't high and doesn't impact potential behavior. It can also be `warning` or `error`, both of which signify a higher impact.

- The descriptive message of what exactly is wrong here: `Unnecessary use of single quotes`.

- The actual lint rule name that is specified in the `analysis_options.yaml`, which is `prefer_double_quotes`.

- Finally, the exact location of the issue is indicated, so that you can fix it.

Depending on the rule, the Dart analyzer might also provide you with a fix, making the process even more automated. You can either manually fix the issue to resolve the linting problem or run `dart fix --apply` to automatically fix all lints that have available fixes (or you can use the IDE counterpart of the CLI command). You can also use other flags with `dart fix`. For example, you can use `dart fix --dry-run` to preview the proposed changes without applying them.

It is important to note that not all lints have automatic fixes, although many of them do, making this tool very useful. The reason why not all of the lints are automatically fixable is that many of them depend on the context of the code and can lead to potential bugs, massive refactoring, and unknown behavior. For example, consider this very simple code: `String getName() => null;`. It will raise a warning that we can't return a `null` where a `String` is expected. However, what would be a correct automatic fix? Should we make the return type nullable? Should we change the return value? Would it be better to implement some logic? It depends, and it's up to the developer to decide, not to the analyzer.

Now, let's explore some strategies for creating our own linting policy.

Exploring lint setup strategies

Many teams and companies that work with Flutter have embraced static code analysis as a tool that greatly contributes to productivity and code quality. This is why many of them have created their own curated lists of lints and made them available to the community. If you are just starting out, or if you trust the judgment and experience of a specific team, you can use their lints as a package. Additionally, there are many advanced tools that create custom lints and static analysis tools. You can research and find what works best for you. For now, let's stick to the decision to configure the lint rules ourselves.

There are two opposite strategies: enabling all lint rules by default or enabling only the ones you need. Let's explore both approaches.

Enabling all lints by default

The list of available lints is quite extensive and new lint rules are occasionally added to it. Going through all of them in a timely manner and not only enabling the ones you want but also configuring them to fit your requirements can be a tedious task. One way to address this, as well as to maintain strict static analysis, is to enable all lints by default and explicitly disable the ones that do not align with your needs. When disabling a lint, it is recommended to leave a comment explaining why it was disabled. This approach makes the linting policy more explicit and easier to understand. Here is an example of how to implement this:

1. Create a new file alongside `analysis_options.yaml` and call it `all_lints.yaml`.

2. Add all of the available rules to this file. You can find them in the official Dart docs at `https://dart.dev/tools/linter-rules/all`.

3. In your `analysis_options.yaml`, change the `include` line to `include: all_lints.yaml` instead of `include: package:flutter_lints/flutter.yaml`.

4. Run the analyzer.

In addition to hundreds of new hints, we will encounter warnings such as "Warning: rule X is incompatible with rule Y." For instance, this occurs with a rule that we have already used: `prefer_double_quotes`. It is incompatible with `prefer_single_quotes` since they aim to accomplish opposite tasks, namely wrapping strings in different types of quotes. This is where personal and team preferences come into play and it becomes your decision to choose one according to your preferences. To avoid these conflicts, we should disable conflicting rules, as shown in the example that follows, in our `analysis_options.yaml`:

```
include: all_lints.yaml

linter:
  rules:
    prefer_double_quotes: false
```

What we did here is identical to what was done before, except this time we set the value of `prefer_double_quotes` to `false`. Since we have now included `all_lints.yaml` instead of `flutter_lints`, a greater number of rules are reporting warnings. Similarly, we should disable any conflicting rules, as well as any specific ones that do not make sense for our team or project.

The opposite strategy does not require additional setup. All you need to do is add the lint rules that you want to your `analysis_options.yaml` file without including all of the lints. However, this approach will require additional research and constant monitoring for new rules as they become available. Both options are viable, although I find the first one to be more elegant and strict, as well as easier to maintain.

Another tip for strictness is what I call a **zero-hint policy**. This means that your code should be structured in such a way that there are zero lint warnings, even at the hint level. You can configure this in the CI/CD pipelines to avoid any debates and keep your lint report log clear. Let's see what else we can do with the analyzer.

Customizing analyzer behavior

In addition to enabling or disabling available rules, we can customize the behavior of the analyzer in various ways. For instance, our `analysis_options.yaml` file can be configured as follows:

```
include: all_lints.yaml

analyzer:
  language:
    strict-casts: true
  exclude:
    - test/_data/**
  errors:
    dead_code: error

linter:
  rules:
    prefer_double_quotes: false
```

Under the `analyzer` section, we have two new subsections. One of them is `exclude`, wherein we can specify the files and folders that we want to exclude from static code analysis. For example, we can exclude the `test` folder. We can also modify the severity of specific rules. Now, the `dead_code` rule will be highlighted as an error instead of just a hint. Finally, in the `language` section, we can enable stricter type checks, such as by setting `strict-casts` to `true`. Let's see what it does. Consider this code:

```
void main() {
  dynamic jsonResponse = '["cupcake", "donut", "eclair"]'; // #1
  List<String> candiesList = jsonDecode(jsonResponse); // #2
  print(candiesList);
}
```

First, we create a variable of the `dynamic` type called `jsonResponse` and assign the value of a `String` type to it, which is `'["cupcake", "donut", "eclair"]'` to imitate a JSON response from an API.

Then we decode that API response with the `jsonDecode` function and assign it to a `candiesList` variable of the `List<String>` type. This code compiles, but once we run it, we will get a runtime error: **Unhandled exception: type 'List<dynamic>' is not a subtype of type 'List<String>'**. That's not good. This is because the `jsonDecode` function return type is `dynamic` and we implicitly cast it to the type List<String>, which it's actually not, because it's List<dynamic>. We can avoid these potential issues by enabling the aforementioned strict-casts. When we do that, we get two error warnings for this code: **error: The argument type 'dynamic' can't be assigned to the parameter type 'String'** and **error: A value of type 'dynamic' can't be assigned to a variable of type 'List<String>'**. We can fix them by adding type checks, such as the following:

```
void main() {
dynamic jsonResponse = '["cupcake", "donut", "eclair"]';
  if (jsonResponse is String) {
     final decodedResponse = jsonDecode(jsonResponse);
     if (decodedResponse is List) {
       final candiesList = decodedResponse;
       print(candiesList.length);
     }
  }
}
```

We have introduced several type checks. Yes, this code is more verbose and requires more boilerplate, but I believe that in a battle between bugs and boilerplate, the latter should win. Catching errors at the compilation stage is always preferable as opposed to catching them during runtime.

Regardless of the strategy you choose, it is important to focus on lints that prevent potential bugs. For instance, we previously discussed a rule in *Chapter 3* that addresses accessing BuildContext across async gaps. I recommended enabling the use_build_context_synchronously lint rule. Now you know how to do it!

However, what if you have a specific use case or style guide that is not covered by the official set of rules? Let's explore several third-party options that are available.

Using the DCM tool

DCM (https://dcm.dev/) is a code quality tool that offers several useful features. First, it adds more than 300 lint rules on top of the already existing Dart and Flutter lints. There are many additional rules that are specific to popular libraries, such as flutter_bloc (https://pub.dev/packages/flutter_bloc), riverpod (https://pub.dev/packages/riverpod), and flame (https://pub.dev/packages/flame), to name a few. Many of the rules offer flexible configurations that allow you to really adjust them to your team preferences. On top of helping with regular lint rules, it can also help you with finding unused or duplicated code and unused dependencies, analyzing the quality of your widgets, and other code metrics such as cyclomatic complexity.

If you need something very specific to your project, or if you're developing a library, you might run into a situation where none of the existing lint rules are enough and you would like to create your own. So let's briefly explore how you can create your own custom lints!

Creating custom lints

To create your own custom lints, you need to create a plugin for the analyzer. However, there are several potential issues with this approach.

First, the task itself is quite cumbersome and requires a good understanding of the API.

Second, the whole feature is currently in an experimental stage (as of 08/25/2024, see `https://dart.dev/tools/analysis#plugins`), which means that there are a few limitations:

- You can only enable one plugin per `analysis_options.yaml` file. So, if you want to create your own plugin and use someone else's, you'll need to choose one.

- It is not recommended to use an analyzer plugin if your development machine has less than 16 GB of memory.

- It is not recommended to use an analyzer plugin if your project structure is a mono-repo with more than 10 packages (in other words, more than 10 `pubspec.yaml` files).

All of the limitations are related to the analyzer requiring a lot of memory. Hopefully, this is something that will be fixed in the future. However, we still have options! In fact, there is a great package that can help us write custom lints without much hassle and will cover the majority of cases. Not everyone needs custom lints, but it's good to know that such an option exists. To achieve this, you can use the `custom_lint` package (`https://pub.dev/packages/custom_lint`). While I won't provide a step-by-step tutorial in this chapter on how to use this library because APIs can change and because this is highly specific, here are some things you should know about using this library:

- Typically, to create a rule, you will extend a class called `DartLintRule` and optionally provide it with a `DartFix`.

- You can have as many packages that implement the `custom_lint` rules in your project as you want. The tool merges them under the hood, combining them into a single analyzer plugin, so you don't need to worry about compatibility issues.

- You can debug your lint rules with logs and write automated tests for them.

This tool is primarily useful for developers working on libraries. However, if your team has highly specific requirements for your code base, it can greatly enhance productivity.

Static analysis is a powerful tool that should be utilized to its fullest extent. However, it is not omnipotent and bugs can still occur during runtime. Let's explore our options for identifying these bugs and determining the causes of other potential issues.

Exploring debugging practices and tools

There are many ongoing debates among programmers. One of them is the print statement versus using debuggers debate. Personally, I often dislike this binary choice because, as is often the case with programming, the only real valid answer is "it depends!" Fortunately, Flutter supports both approaches and provides us with some amazing tools. Let's review them.

Logging – the good, the bad, and the ugly

The logging process is a widely used method to obtain information about the behavior of our app, which is printed into the console. Most programming languages, including Dart, have a method for printing statements to the console. In Dart, this is done by calling the `print('Any string here');` method. However, this simple approach has limitations when used in the context of an application. The following issues arise when using the `print` function:

- Very long log messages can be truncated, which is especially inconvenient when logging network communication
- Messages may be dropped if there are too many of them at the same time

These problems can be easily solved by using a Flutter-specific method called `debugPrint`:

- It has an algorithm to handle message truncation
- It is better at handling throttling and can be customized to fit specific needs

So, the first step would be to use `debugPrint` instead of `print`. However, for a production-level application, even that is not sufficient. Typically, we would want the following:

- Access to different levels of log severity, such as `debug`, `warning`, and `error`. It would also be helpful to filter these logs based on the build type. For example, you might want to show all logs for DEV builds, show warning and error logs for QA builds, and only show error logs for RELEASE builds.
- Instead of displaying logs to the console, it would be beneficial to output them to an error reporting system of our choice. This allows us to analyze and address any issues in production promptly.

Although there is no built-in support for such a mechanism, you still have two options. I highly recommend using one of them:

- There are many well-maintained and feature-rich loggers created by the community. Research the available options and choose the one that best suits your needs.
- Alternatively, you could create your own wrapper around the default tools.

Use logging wisely and always remember not to log sensitive user data. Be mindful of your logs and who has access to them. What else can we do to catch potential problems in debug mode?

Using assertions to catch errors in debug mode

Another useful tool provided by Dart is called **assertions**. It is important to note that assertions are only available in debug mode. This is because they come with a runtime cost and are designed to halt app execution when a runtime error occurs. This is not something we would want in production. However, assertions can provide an additional layer of safety during development time. Let's consider the `CartState` constructor as an example and see how we can improve it using assertions:

```
const CartState({
    required this.items,
    required this.totalPrice,
    required this.totalItems,
    required this.loadingResult,
  })   : assert(totalPrice >= 0),
         assert(totalItems >= 0);
```

What we did here is add two `assert` statements to our `CartState` constructor. These statements check the values of the `totalPrice` and `totalItems` variables and ensure that they are positive, as negative values do not make logical sense in this case. If something goes wrong with our code and we pass inappropriate data to the `CartState` constructor during debugging, it will throw an error and require us to handle it in the code before it reaches production. Nice. However, what if assertions don't help and our app starts behaving strangely? Besides logging, what options do we have to inspect the control flow of our code?

Debugging code with breakpoints

So, you have logged your way through the application and noticed that your code is entering a condition that it shouldn't. Why is it doing so? How did it end up there? To answer these questions, you can use breakpoints. All modern IDEs support debugging with breakpoints in Flutter. You can find the specific usage instructions for your preferred IDE. If you're unfamiliar with breakpoint debugging, here's how it works:

1. In your IDE, select the line of code that is causing concern and set a breakpoint using the provided tools.

2. This means that when you run your app in debugging mode, the code will pause once it reaches this line, as if taking a break.

3. At this point, the IDE will offer various tools:

    *   Review the call stack that led to this code.

    *   Inspect the values of all variables in this context.

- Evaluate expressions in this context.
- Continue code execution step by step, either by going deeper with **step into** or going farther with **step over** and **step out**.

Breakpoints are powerful tools that can be confusing, especially when you're just starting out. However, they are essential for locating and fixing tricky bugs. Combining breakpoints with logging can yield the best results quickly.

That's not all. Flutter provides us with another incredibly powerful tool, or rather a set of tools, called Flutter DevTools.

Diving into the world of Flutter DevTools

Flutter DevTools is a comprehensive suite of tools that helps developers investigate various aspects of their app, from UI performance to network monitoring. Exploring all the features of this suite would require a full chapter or even a whole book, but that wouldn't be practical considering the visual nature of these tools and the ever-changing nature of software, particularly in terms of graphics. It's better to keep up with the latest updates rather than relying solely on a book. However, we will still review what can be done with these tools and how they can enhance the quality of our app.

To launch Flutter DevTools, you would typically click on a designated button in your preferred IDE and run your application in either debug or profile build mode. In the Flutter DevTools console, you have the following views:

View	Functions
Flutter Inspector	• Inspect the widget tree visually • View widget properties and details, as well as change some of them in runtime • Debug layout issues • Debug performance issues by slowing down animations and highlighting oversized images • Highlight widget repaints
Performance view	• Analyze a timeline of frame-by-frame performance • Get details on shader compilation jank • Inspect raster stats • Track widgets builds, layouts, and paints

View	Functions
CPU profiler view	• Profile CPU performance • Inspect method call stacks to identify expensive paths • Inspect method table statistics
Memory view	• Profile memory usage • Find memory leaks
Network view	• Inspect HTTP, HTTPS, and web socket traffic
App size tool	• Analyze the total size of your app • Inspect what code causes increases in app size

Table 1.1 – Overview of Flutter DevTools views

As you can see from the table, there are many things that can go wrong and that can require inspection. However, we have amazing tools that are very helpful in our debugging journey. Make sure to practice using all of them so that once you find yourself in a situation that requires some serious debugging, you will know exactly what to do.

Summary

In this chapter, we learned why maintaining code consistency is an important matter and overviewed a lot of practices and tools that make our code more clean, idiomatic, and reliable. We learned how to automate code formatting and leverage the power of static code analysis with the help of the Dart formatter, Dart analyzer, and custom_lints package. We explored various approaches, including their strengths and limitations, as well as toolsets for quickly locating and fixing any potential issues in our apps with loggers, breakpoints, and the Flutter DevTools.

With that, we have arrived at the end of our journey. Now you are well equipped with a toolset of design patterns and best practices applicable for Flutter applications. You have also gained a deeper understanding of the inner workings of the Flutter framework, which influence the selection of those patterns. It is important to remember that patterns are just guidelines and you should always be mindful of the use case at hand in order to analyze the problem and choose the best solution, as opposed to blindly following a proposal. I hope that this book has helped you develop a more critical attitude toward software development and has inspired you to apply best practices and design patterns to create even more fantastic Flutter applications. Happy building!

Get this book's PDF version and more

Scan the QR code (or go to `packtpub.com/unlock`). Search for this book by name, confirm the edition, and then follow the steps on the page.

Note: Keep your invoice handy. Purchases made directly from Packt don't require an invoice.

13
Unlock Your Exclusive Benefits

Your copy of this book includes the following exclusive benefits:

DRM-Free PDF Version

Download DRM-free PDF and ePub copies of this book.

7-Day Packt Library Access

Get 7-day unlimited access to 8,000+ books and videos. No credit card required.

Available for first-time Packt+ trial users only.

Next-Gen Reader Access

Read this book on Packt Reader with progress sync, dark mode and note-taking.

Follow the guide below to unlock them. The process takes only a few minutes and needs to be completed once.

Unlock this Book's Free Benefits in 3 Easy Steps

Step 1

Keep your purchase invoice ready for *Step 3*. If you have a physical copy, scan it using your phone and save it as a PDF, JPG, or PNG.

For more help on finding your invoice, visit `https://www.packtpub.com/en-us/unlock?step=1`.

> **Note:**
> If you bought this book directly from Packt, no invoice is required. After *Step 2*, you can access your exclusive content right away.

Step 2

Scan the QR code or go to `packtpub.com/unlock`.

On the page that opens (similar to *Figure 13.1* on desktop), search for this book by name and select the correct edition.

Unlock Your Book's Free Benefits

Bought a Packt book from Amazon or one of our channel partners? Unlock your free benefits in 3 easy steps.

Figure 13.1: Packt unlock landing page on desktop

Step 3

After selecting your book, sign in to your Packt account or create one for free. Then upload your invoice (PDF, PNG, or JPG, up to 10 MB). Follow the on-screen instructions to finish the process.

Need Help

If you get stuck and need help, visit https://www.packtpub.com/unlock-benefits/help for a detailed FAQ on how to find your invoices and more. This QR code will take you to the help page.

> **Note:**
> If you are still facing issues, reach out to customercare@packt.com.

Index

www.packtpub.com

Subscribe to our online digital library for full access to over 7,000 books and videos, as well as industry leading tools to help you plan your personal development and advance your career. For more information, please visit our website.

Why subscribe?

- Spend less time learning and more time coding with practical eBooks and Videos from over 4,000 industry professionals
- Improve your learning with Skill Plans built especially for you
- Get a free eBook or video every month
- Fully searchable for easy access to vital information
- Copy and paste, print, and bookmark content

Did you know that Packt offers eBook versions of every book published, with PDF and ePub files available? You can upgrade to the eBook version at packtpub.com and as a print book customer, you are entitled to a discount on the eBook copy. Get in touch with us at customercare@packtpub.com for more details.

At www.packtpub.com, you can also read a collection of free technical articles, sign up for a range of free newsletters, and receive exclusive discounts and offers on Packt books and eBooks.

Other Books You May Enjoy

If you enjoyed this book, you may be interested in these other books by Packt:

Flutter Cookbook, Second Edition

Simone Alessandria

ISBN: 978-1-80324-543-0

- Familiarize yourself with Dart fundamentals and set up your development environment
- Efficiently track and eliminate code errors with proper tools
- Create various screens using multiple widgets to effectively manage data
- Craft interactive and responsive apps by incorporating routing, page navigation, and input field text reading
- Design and implement a reusable architecture suitable for any app
- Maintain control of your codebase through automated testing and developer tooling
- Develop engaging animations using the necessary tools
- Enhance your apps with ML features using Firebase MLKit and TensorFlow Lite
- Successfully publish your app on the Google Play Store and the Apple App Store

Flutter for Beginners

Thomas Bailey, Alessandro Biessek

ISBN: 978-1-83763-038-7

- Understand the Flutter framework and cross-platform development
- Acclimate the fundamentals of the Dart programming language
- Explore Flutter widgets, the core widget library, and stateful and stateless widgets
- Discover the complete development lifecycle, including testing and debugging
- Get familiar with both the mobile and web app release processes
- Dig deeper into more advanced Flutter concepts like animation
- Explore common Flutter plugins ad how to use them
- Discover the Flutter community and how to stay up-to-date

Packt is searching for authors like you

If you're interested in becoming an author for Packt, please visit authors.packtpub.com and apply today. We have worked with thousands of developers and tech professionals, just like you, to help them share their insight with the global tech community. You can make a general application, apply for a specific hot topic that we are recruiting an author for, or submit your own idea.

Share Your Thoughts

Now you've finished *Flutter Design Patterns and Best Practices*, we'd love to hear your thoughts! Scan the QR code below to go straight to the Amazon review page for this book and share your feedback or leave a review on the site that you purchased it from.

https://packt.link/r/1-801-07264-7

Your review is important to us and the tech community and will help us make sure we're delivering excellent quality content.

www.ingramcontent.com/pod-product-compliance
Lightning Source LLC
Chambersburg PA
CBHW080614060326
40690CB00021B/4695